Schools, Mathematics, and the World of Reality

Edited by

ROBERT B. DAVIS
CAROLYN A. MAHER
Rutgers, The State University of New Jersey

ALLYN AND BACON
Boston   London   Toronto   Sydney   Tokyo   Singapore

Copyright © 1993 by Allyn and Bacon
A Division of Simon & Schuster, Inc.
160 Gould Street
Needham Heights, Massachusetts 02194

All rights reserved. No part of the material protected by this copyright notice may be reproduced or utilized in any form or by any means, electronic or mechanical, including photocopying, recording, or by any information storage and retrieval system, without written permission from the copyright owner.

Library of Congress Cataloging-in-Publication Data

Schools, mathematics, and the world of reality / Robert B. Davis and
 Carolyn A. Maher, editors.
   p.   cm.
  Papers originally presented at a conference in the fall of 1990 at
 the Rutgers Graduate School of Education, New Brunswick, N.J.
  Includes bibliographical references and index.
  ISBN 0-205-13445-9
  1. Mathematics—Study and teaching—Congresses I. Davis, Robert
 B. (Robert Benjamin),   . II. Maher, Carolyn Alexander.
 QA11.A1S363   1993
 372.7′044—dc20   92-11678
                                                       CIP

ISBN 0-205-13445-9
H34457

Printed in the United States of America
10 9 8 7 6 5 4 3 2 1   96 95 94 93 92

Contents

Series Foreword   v
    by Louise Cherry Wilkinson

Contributors   viii

Introduction: Schools, Mathematics, and the World of Reality   ix
    by Robert B. Davis and Carolyn A. Maher

PART ONE
Context

Chapter One · *Mathematics Education: The Context of the Crisis*   1
    by Gerald A. Goldin

Chapter Two · *What Are the Issues?*   9
    by Robert B. Davis and Carolyn A. Maher

Chapter Three · *Constructivism and Caring*   35
    by Nel Noddings

PART TWO
Relating Schools to Reality

Chapter Four · *The Reality of Negative Numbers*   51
    by Robert B. Davis and Carolyn A. Maher

Chapter Five · *Learning Mathematics: Perspectives from Everyday Life*   61
    by Terezinha Nunes

Chapter Six · *Understanding Mathematics: The Impact of Technology*   79
    by Anthony Ralston

PART THREE
What Should a Classroom Look Like?

Chapter Seven · *Learning to Use Children's Mathematics Thinking: A Case Study*   93
    by Elizabeth Fennema, Thomas P. Carpenter, Megan L. Franke, and Deborah A. Carey

Chapter Eight · *Mathematics as Procedural Instructions and Mathematics as Meaningful Activity: The Reality of Teaching for Understanding*   119
    by Paul Cobb, Terry Wood, Erna Yackel, and Elizabeth McNeal

Chapter Nine · *Students' Disagreements During Small-Group Mathematical Problem Solving*   135
    by Louise Cherry Wilkinson and Amy Martino

PART FOUR
The Reality of Students

Chapter Ten · *Brian's Representation and Development of Mathematical Knowledge: The First Two Years*   173
    by Carolyn A. Maher, Robert B. Davis, and Alice Alston

Chapter Eleven · *Mathematics and the Reality of the Student: Bringing the Two Together*   213
    by Anna O. Graeber

Chapter Twelve · *Assessing Mathematical Thinking and Learning Potential*   237
    by Herbert P. Ginsburg, Susan F. Jacobs, and Luz S. Lopez

Chapter Thirteen · *Real Tasks and Real Assessment*   263
    by Jan de Lange

Index   288

Series Foreword

Within the past decade, the profession of education has been shaken to its roots as national attention focused on education and educators and on their perceived lack of success with students and their achievement. Critics and friends have raised basic questions about the profession, including whether educational professionals have successfully met the challenges that the students and the schools present and, even more fundamentally, if they are able to meet those challenges. Beginning with the highly publicized *A Nation at Risk,* seemingly endless and often contradictory criticisms, analyses, and recommendations have appeared from virtually every segment of contemporary American society.

Perhaps more than any other content area, mathematics achievement arouses our national concern. American students do not achieve at the same levels as most students from the highly developed countries in the areas of arithmetic, algebra, geometry, and calculus. Additionally, there are severe problems with U.S. students' attitudes and the attitudes of their parents toward mathematics. Most of our graduate schools no longer enroll any U.S. citizens in graduate programs in mathematics. Data from the National Assessment of Educational Progress has revealed that American children seem to master basic computation with whole numbers but are less capable in working with decimals and fractions. The greatest concern is the apparent inability of American children to use these conceptual skills in the real world to solve standard mathematical word problems. Experts in mathematics education believe that the underlying cause is that our students have very little understanding of even those properties that they can perform algorithmically. They are unable to explain why they work, what they represent, and what the standard notations involved stand for. They cannot explain why their answers are reasonable or unreasonable. The conclusion is that these students are unable to reason and communicate mathematically. One of the major causes of this deficit is that they have been exposed to curricula that have stressed procedural knowledge and not conceptual knowledge. More emphasis has been placed on mindless doing than on understanding.

It is not surprising, then, that several important recent documents in mathematics education—"Everybody Counts" from the National Research

Council, and *Curriculum and Evaluation Standards for School Mathematics, Constructivist Views on the Teaching and Learning of Mathematics,* and *Professional Standards for Teaching Mathematics,* all published by the National Council of Teachers of Mathematics — call for the complete overhauling of the way mathematics is taught and learned in the United States. These reform efforts call for reformulating the mathematics curricula and teacher preparation to focus on problem solving and reasoning. The focus would be expressed in objectives for schools that emphasize understanding the meaning of important mathematical concepts, not simply mastering algorithms by rote.

In this recent explosion of concern for educational reform, we see a need for a general and national forum in which the problems of education can be examined in light of research from a range of relevant disciplines. Too often, analyses of very complex issues and problems occur within a single discipline. Aspects of a problem that are unfamiliar to those members of the discipline are ignored, and the resulting analysis is limited in scope and hence unsatisfactory. Furthermore, when educational issues are investigated by members of one discipline alone, there is seldom an attempt to examine related issues from other fields or to apply methods developed in other fields that might prove illuminating.

The national debate on educational reform has suffered from this myopia, as problems and issues are identified and analyses and solutions often are proposed within the limited confines of a single disciplinary boundary. In the past, national discussions have been ill informed or uninformed by current research, partly because there are far too few mechanisms for interdisciplinary analyses of significant issues.

The present series of volumes, the Rutgers Symposium on Education, attempts to address this gap. The series focuses on timely issues and problems in education from an interdisciplinary perspective. The focus of each volume will be a particular problem, such as a potential teacher shortage, the structure of schools, or the effects of cognitive psychology on how to teach mathematics. There is an accumulating body of high-quality educational research on topics of interest to practitioners and policymakers alike. Each volume in the series will provide an interdisciplinary forum through which scholars can disseminate their original research and extend their work to potential applications for practice, including guides for teaching, learning, assessment, intervention, and policy formulation. We believe that this work will increase the potential for significant analysis and for positive impact in the domains of both practice and theory.

The fourth volume in the series presents critical analyses of the mathematics education reform movement in the United States. The editors note:

> Even mathematics itself is nowadays seen in a different light; where it once appeared as a collection of rules to be learned and followed,

it is now seen as a flexible and powerful *way* of thinking, as a matter more of *analytical skill* than of following memorized procedures. Still more fundamentally, we might argue that *children themselves* have changed — perhaps not literally, but as we have come to listen more carefully to children, we have not failed to notice that children invent many more mathematical ideas than were previously expected from them. In that sense, the "modern" child is more inventive, more creative, more thoughtful — one would almost say "smarter" — than we used to believe.

The contribution of the work presented in this book is, without doubt, substantial. All too often, too many rush to the most current movement, seeing it as the panacea for education without really understanding what was done and why it did or did not work. By carefully considering one of the most significant education reform movements in the United States today, the authors contribute new knowledge.

This book is based on original research papers presented by the authors at a conference in the fall of 1990 in New Brunswick, New Jersey, at the Rutgers Graduate School of Education. This conference was cosponsored by the Rutgers Graduate School of Education and the Center for Mathematics, Science, and Computer Education and the National Center for Research in Mathematical Sciences Education and the Wisconsin Center for Education Research (headquartered at the University of Wisconsin-Madison). It is with great pleasure that we contribute this series of volumes on contemporary educational issues, the Rutgers Symposium on Education. Our expectation is that this series will serve as a seminal contribution to the literature in educational theory and practice.

> Louise Cherry Wilkinson
> Dean of the Rutgers Graduate School of Education
> and Professor of Educational Psychology

Contributors

Alice Alston, Rutgers, The State University of New Jersey
Deborah A. Carey, Thomas Carpenter, University of Wisconsin-Madison
Paul Cobb, Purdue University
Robert B. Davis, Rutgers, The State University of New Jersey
Jan de Lange, OW & OC, Utrecht, Holland
Elizabeth Fennema, University of Wisconsin-Madison
Herbert P. Ginsburg, Teachers College, Columbia University
Gerald A. Goldin, Rutgers, The State University of New Jersey
Anna O. Graeber, University of Maryland, College Park
Susan F. Jacobs, Teachers College, Columbia University
Megan L. Franke, University of Wisconsin-Madison
Luz S. Lopez, Teachers College, Columbia University
Carolyn A. Maher, Rutgers, The State University of New Jersey
Amy Martino, Rutgers, The State University of New Jersey
Elizabeth McNeal, Purdue University
Nel Noddings, Stanford University
Terezinha Nunes, University of London
Anthony Ralston, State University of New York-Buffalo
Louise Cherry Wilkinson, Rutgers, The State University of New Jersey
Terry Wood, Purdue University
Erna Yackel, Purdue University

Introduction: Schools, Mathematics, and the World of Reality

ROBERT B. DAVIS   CAROLYN A. MAHER

The title of this book requires explanation. Has it not always been a goal to have our schools related as closely as possible to the real needs, values, and possibilities of our society?

The problem, of course, is that society outside of school changes and, nowadays, changes very rapidly indeed. A school curriculum and methods of promoting learning that were designed for the early twentieth century are far less well adapted to the approach of the twenty-first century.

These changes exist in every aspect of school. Most obvious is the change in technology, and especially the advent of computers. Not in existence when today's schools were designed, computers today are ubiquitous; nearly every automobile, television set, microwave oven, and dishwasher made now contains some kind of computer, although the purchaser may not be aware of this. (Of course, microwaves, televisions, and dishwashers themselves did not exist when today's school programs were devised.) Chapter 6 by Ralston looks at this development explicitly, and other chapters do in less direct ways. (So fast is the pace of change that Ralston finds it necessary to distinguish between modern, state-of-the-art computer programs and programs created *in the early days of computers!*)

But technology is by no means the only difference. Modern cognitive psychology has, fortunately, freed us from the stimulus–response school of thought and allowed us to begin to take account of thought itself. After all, what is most special about human beings, who do not closely resemble pigeons or rats? Surely the answer is the distinctive ways that humans think, feel, and care, and nowadays we need to attend far more carefully to what is truly human. Noddings (Chapter 3) has long been known as a leader in this field, and several other chapters also deal with this matter. In particular, the modern emphasis on *thought* has given us the *constructivist* approach to mathematics education, which permeates this book, as it could hardly fail to do, given that several authors are leading creators of this movement.

We begin to see, as in the study of Brian (whom we meet in Chapter 10), the complexity of the thought processes of a child who works at building meaningful interpretations of situations. The analyses made possible by studying the videotapes of Brian doing mathematics provide us with more insight into how he builds up representations of mathematical ideas. From the perspective of the practicing teacher, this level of detailed analysis is not usually a realistic possibility. However, Fennema, Carpenter, Franke, and Carey (Chapter 7) document how a teacher changed her teaching as she came to pay greater attention to the thought processes of her students.

Indeed, it is no exaggeration to speak of "modern" pedagogical methods as opposed to "traditional" ones. Schools used to be thought of in terms of information transfer. Teachers (and adults generally) knew things that children did not know, and this knowledge had to be told to children. "Telling" was a major pedagogical method, perhaps *the* major pedagogical method. Modern pedagogy, by contrast, sees schools as places to help children develop their own thinking. We need to find yet other ways to help teachers pay closer attention to children's thinking. De Lange (Chapter 13) challenges us to provide an alternative to evaluation in its traditional form — that is, to use more "open-ended" tasks that make possible a meaningful assessment of mathematics. Such an approach to assessment complements pedagogical methods that support the development of thinking in children and a shift in the role of the teacher to a coach — one who helps each student to see what he or she is doing, and who helps each student find ways to do even better.

Even mathematics itself is nowadays seen in a different light. Where it once appeared as a collection of rules to be learned and followed, it is now seen as a flexible and powerful *way* of thinking, a matter more of *analytical skill* than of following memorized procedures.

Still more fundamentally, we might argue that *children themselves* have changed — perhaps not literally, but as we have come to listen more carefully to children, we have not failed to notice that children invent many more mathematical ideas than were previously expected from them. In that sense, the "modern" child is more inventive, more creative, more thoughtful — one would almost say "smarter" — than we used to believe.

Terezinha Nunes's studies of Brazilian children and adults, as discussed in Chapter 5, underscore the power of learning mathematics from situations that one deals with in everyday life. Schools, of course, are a reality in themselves, and sometimes their development is guided by an inner dynamics that may lead them out of touch with the external world. In this way a subject has grown up that is sometimes called "school mathematics," which is often quite different from mathematics. Textbooks present problems such as

"Express 4 as a percent," a task which stumps both mathematicians and business leaders, who would not consider this to be a well-defined task (see, e.g., Davis, 1988; Brown, Collins, & Duguid, 1989; Elsholz & Elsholz, 1989; Davis, 1989). The mathematical ideas in the mind of a student may differ from all of these (see, e.g., Erlwanger, 1973; Rosnick & Clement, 1980; McNeill, 1988); indeed, we are coming to realize that misconceptions are almost inevitable as a student builds up new ideas. In Chapter 11, Anna Graeber explores such differences in a careful and scholarly way, describing (among many other things) the harshly conflicting pressures that confront schools from, on the one hand, the "back-to-basics" advocates and, on the opposite side, those who argue that mathematics is supposed to be a *thoughtful* subject and not mere rote. Especially important is her recognition of one of the greatest realities of all, the reality of the child as an individual person who thinks and has ideas—ideas that, even when well founded in the child's limited experience, may be wrong and may also be highly resistant to change. This aspect of mathematics teaching was hardly recognized until the last few decades.

One of the most prominent proposals of recent years has been the idea of students working together in small groups. This arrangement may have had its modern origins in England; see, for example, Marshall (1963), Ridgway and Lawton (1965), Blackie (1971), and Weber (1971). It came to the United States partly through the efforts of the Educational Development Center and probably primarily through schools in California, plus some influential private schools in New York City (see, e.g., Featherstone, 1971; Rathbone, 1971; Silberman, 1973, Devaney, 1974; Davidson, 1990). But a classroom that is not dominated by the teacher, a classroom that instead depends on student cooperation in small groups, can be a very new and very threatening reality. What goes on in all those small groups, anyhow? This question is addressed in Chapter 9 by Louise Cherry Wilkinson and Amy Martino.

There are at least two other realities that need discussion: first, the question of how well our schools are preparing our young people to become productive, responsible citizens in a democracy and, second, whether the research methods currently used by scholars to describe and analyze the process of schooling are reasonably adequate to the task. These questions are considered in Chapter 2 by Davis and Maher, entitled "What Are the Issues?"

These and many other themes were considered during a two-day conference held at Rutgers University November 15–16, 1990, and are discussed in these chapters. We do not try to anticipate them in detail, but we hope that readers will find a new view of schools, children, mathematics, and the process of teaching mathematics as they read the various chapters in this book.

This book is, we hope, much more than "just one more book about constructivism." It says: "Look at today's child from the perspective of today's psychology. Look at today's world, and think of the prospects for the twenty-first century in which this child will live and work. Look at today's ways of creating organizations and getting jobs done, from hospitals to factories to farms to insurance companies. Now ask yourself: Are our schools designed as we design our modern organizations? Are they designed in order to maximize the opportunities for our young people to grow and mature into the kind of adults today's democracy needs as citizens? Is it the twenty-first century that we seem to be aiming for?"

We felt that the answer to the questions had to be a resounding "No!" That is why we held the conference that gave rise to this book. There is much that needs to be done, but there is a growing theoretical foundation on which new approaches can be created, and much of this foundation is suggested in the chapters that follow.

References

Blackie, J. (1971). *Inside the primary school.* New York: Schocken.
Brown, J. S., Collins, A., & Duguid, P. (1989). Situated cognition and the culture of learning. *Educational Researcher, 18*(1), 32–42.
Davidson, N. (Ed.). (1990). *Cooperative learning in mathematics.* Menlo Park, CA: Addison-Wesley.
Davis, R. B. (1988). Is percent a number? *Journal of Mathematical Behavior, 7*(3), 299–302.
Davis, R. B. (1989). The culture of mathematics and the culture of schools. *Journal of Mathematical Behavior, 8*(2), 143–160.
Devaney, K. (1974). *Developing open education in America.* Washington, DC: National Association for the Education of Young Children.
Elsholz, R., & Elsholz, E. (1989). The writing process: A model for problem solving. *Journal of Mathematical Behavior, 8*(2), 161–166.
Erlwanger, S. H. (1973). Benny's conception of rules and answers in IPI mathematics. *Journal of Children's Mathematical Behavior, 1*(2), 7–26.
Featherstone, J. (1971). *Schools where children learn.* New York: Liveright.
Marshall, S. (1963). *An experiment in education.* London: Cambridge University Press.
McNeill, R. (1988). A reflection on when I loved math and how I stopped. *Journal of Mathematical Behavior, 7*(1), 45–50.
Rathbone, C. (Ed.). (1971). *Open education: The informal classroom.* New York: Citation.
Rosnick, P., & Clement, J. (1980). Learning without understanding: The effect of tutoring strategies on algebra misconceptions. *Journal of Mathematical Behavior, 3*(1), 3–27.
Silberman, C. (Ed.). (1973). *The open classroom reader.* New York: Vintage.
Weber, L. (1971). *The English infant school and informal education.* Englewood Cliffs, NJ: Prentice-Hall.

CHAPTER ONE

Mathematics Education
The Context of the Crisis

GERALD A. GOLDIN

In my recent travels I have had conversations with many people who are deeply concerned about mathematics education. This introductory chapter reflects the perspective at which I have arrived on the situation nationally and locally, both from these discussions and from my work directing the Center for Mathematics, Science, and Computer Education at Rutgers University.

The social context in which we are now striving for reform in mathematics education is, unfortunately, quite negative, and the trend offers little reason for optimism. Several years have passed since the president and the Congress of the United States ostensibly awoke to the danger of our country falling behind in mathematics and science. In this time a lot of discussion has occurred, but there has been almost no movement toward fundamental, structural change in our educational system. The system is deteriorating rather than improving.

In many communities, education still takes place in a context where children have little material, educational, or emotional support at home. Schools remain unable to provide the support that is lacking. In many states there remain vast discrepancies between education in poorer and in more affluent school districts; in some (including New Jersey), where financial support is being reallocated from wealthier districts to poorer ones, it is not clear that the funding will find its way into programs that make a difference.

At all socioeconomic levels, our society persists in setting low educational goals. In wealthy, suburban communities, where the intellectual and physical resources for high-quality education are generally available, there is a disturbing tendency for schools to coast, particularly in mathematics and science. Here it is easy for school administrators to cite high achievement levels, evidenced by standardized test scores and student admissions to prestigious universities, as hallmarks of their schools' successes—although these may be due more to the high socioeconomic status of parents than to high-quality education. Why push our children, when they are already doing fine? Why pursue advanced goals in cooperation with universities, when the schools are perfectly capable of doing the job themselves? Why take risks, when bureaucracy and politics reward stability and predictability? In less affluent communities the prevailing objectives are more modest still. Our society has internalized—and now takes for granted—the cruel myth that minority and economically deprived children (with rare exceptions) cannot succeed in school. Despite the human cost in lost potential, it is widely accepted as sufficient if, in poor urban and rural communities, scores on minimal proficiency tests can be somewhat improved. The possibility of achieving excellent science and mathematics education for all students in the United States is regarded as unrealistic, and not to be taken seriously.

These widespread attitudes reflect a persistent national complacency and denial in the face of several related crises: our enormous budget deficit, our squandering of precious national resources (e.g., in the savings and loan disaster, the Housing and Urban Development scandal, and massive commitments to new military objectives despite the end of the Cold War), our declining industrial competitiveness—and, of course, our unwillingness to develop an educational system that meets the present and future needs of our children. As a society, we have learned to behave as if our actions have no consequences.

With regard to mathematics education, public apathy remains despite the valiant efforts of major national groups, such as the National Council of Teachers of Mathematics (NCTM) and the Mathematical Sciences Education Board, that have sought to offer a vision of what is necessary—and possible—to achieve. When as professionals in the field of education we plan strategies for improving the teaching of mathematics, we must take these aspects of the social context into account. In "relating mathematics to the real world," we must consider the real world in which schools exist and children live, a world in which academic achievement is undervalued by the students' peers as well as by the wider society. We must find ways to give mathematics meaning, value, and excitement for students in their lives as they actually are. Because we do not yet have the material resources for broad societal change, we must create smaller scale working models of the ideal programs we would like to see. Somehow, too, we must achieve a far more dynamic commitment within our society to positive change in the field of education.

What are some of the changes that are most important from the perspective of the teaching of mathematics?

First, we must not only continue developing our understanding of how children learn mathematics effectively, but we must implement the knowledge we already have. There is already a wide consensus among researchers and leading mathematics teachers that powerful mathematics learning cannot be achieved through traditional rote methods. Most children do not develop conceptual understanding when they are presented with low-level, discrete skills and algorithmic procedures to imitate, procedures in which they are then given "drill and practice" and on which they are subsequently tested. Furthermore, most children learn to dislike mathematics when it is taught this way. Such a model simply does not work. Current research directs us toward far more sophisticated, cognitively oriented approaches: approaches based on the construction of meaning for mathematical ideas, discovery learning and problem solving, the building of powerful systems of cognitive representation, and the provision of structured learning environments that foster such outcomes.

Nevertheless, our schools have been generally unable to move beyond the rote model. One reason is that the assessment systems now in place are still largely based on such a model. Our orientation toward mathematics is such that in virtually every school district it is clear (at all levels from the superintendent down) that the goal is to raise test scores. Parents and school board members accept this as the bottom line. And the seemingly logical strategy, once this goal is asserted, is to perform a skills analysis of what is needed to answer the test items correctly, to decompose the skills further into procedures to be taught each week and each day, and to establish the resulting unit and lesson plans as the school curriculum. On the basis of a certain means–ends interpretation of what doing mathematics is, this seems like good planning—except that it leaves no room for the ways in which children really do learn mathematics meaningfully. There is growing evidence that this strategy, adopted nationwide to improve mathematics achievement, is failing, even according to the traditional definition of achievement: Test scores have leveled off at mediocre values and remain there year after year. For students to go beyond one- or two-step problems in mathematics requires conceptual understanding, not the ability to perform memorized operations in sequence; in removing the development of this understanding from the curriculum, we have removed the foundation on which mathematics is built. We wind up teaching and reteaching the same arithmetic skills, year after year, until at the university level our students still do not have them.

A second important domain of needed change is that of teacher enhancement. This embraces two related aspects—the role of the teacher as a professional, and the expectations we have for mathematics teachers as knowledgeable practitioners.

It is crucial that we soon come to the point of recognizing teachers in our society as real professionals, commanding a specialized body of valuable knowledge, who can collaborate with experts from other domains pertaining to education in turning our schools around. But in many school districts the teacher has little autonomy or decision-making authority. Consequently, teachers with the most original or creative ideas often find themselves stifled and hamstrung. Some develop mechanisms of coping with the bureaucratic context—filling out necessary forms, nodding to authority, closing the classroom door, and doing things their own way despite it all. Others cannot cope this way, and many of our best teachers become alienated from their profession.

There is an admirable undercurrent of idealism in our country, which has taken us very far. One expression of this is through groups such as Teach for America, which involves young, energetic college students in making a commitment to the schools, in giving back something of value from the education they received. What will happen to this creative energy? Will these young people, after some real school experience, come to love teaching and want to participate on a long-term basis in our schools? Or will they be disillusioned by a system that can ultimately make no use of their talent or commitment; will they serve simply as temporary, inexpensive replacements for older teachers who have been "burned out"? The answer depends on whether there is a national commitment to develop teachers who really are professionals, with the capacity to implement change as professionals. This commitment has not yet been made.

A related issue is that of cultural barriers in the mathematical and scientific professions. Our school systems themselves impose a rigid compartmentalization between elementary and secondary education, with little interaction among teachers at different levels. There are still higher barriers between K–12 schools and colleges; these are two different communities, scarcely interacting except when college teachers visit schools as parents, or teachers take education courses at the university. Both educational communities are separate from industry: In New Jersey, corporate research laboratories employ mathematicians, scientists, and statisticians in large numbers, but have little interaction with schools.

In emphasizing this I do not mean to belittle the efforts of those who are trying to change the situation. The Center at Rutgers sponsors many ongoing projects involving elementary and secondary school teachers, university faculty, and industry scientists, as do similar centers at other colleges and universities. AT&T Bell Laboratories, Merck, Bellcore, Johnson & Johnson, Exxon, and other corporations have ongoing programs, and the New Jersey Business and Industry Science Education Consortium (BISEC) sponsors summer research by teachers in industrial laboratories. New Jersey is one of the national leaders in overcoming existing professional barriers. Yet, as a physicist, I am used to estimating orders of magnitude through what we call back-of-the-envelope calculations, and my best estimate is that in New Jersey we are currently committing toward innovation less than one percent of the resources that would be needed to make a genuine change in the situation. The initiatives that have been undertaken must be increased drastically if we are to arrive at a new cultural context—one in which elementary and secondary teachers have seen in some depth pure and applied mathematical research, and move easily in the university and in industry; one in which research mathematicians and scientists know some of the problems of education, and move easily in schools; a context based on one large community of mathematical and scientific researchers and educators, rather than the disjointed groups we have now.

As we seek to raise mathematics teachers' professional status, we must also expand our expectations for their knowledge base. One component of that expertise, implicit in what I have already advocated, has to do with experience in the activity of doing mathematics. We have failed to provide that experience through traditional expository mathematics courses in universities. It is important that we create something different—learning opportunities in mathematics that not only address modern, exciting topics, but do so in an involving way. We must model at the university level the kinds of creative, discovery-oriented learning experiences we would like to see in elementary and secondary classrooms. Several fields of research mathematics not only are quite current, but also offer concepts that are accessible to teachers and students alike: some domains of discrete mathematics, fractal geometry, the study of chaos, and so on. These are ideal choices for development. The idea that teachers are unable to keep up with modern research must be dispelled as an unproductive (though unfortunately self-fulfilling) preconception.

A second component of the knowledge base for professional mathematics teachers that needs enhancement is the effective modeling of the student's mathematical learning. The capable teacher must not only understand the mathematics, but also understand the student's understanding well enough to provide for its powerful development over time. This is a difficult task, but it is one that many teachers perform very effectively when they are not hemmed in by the requirement of teaching toward skills-oriented standardized tests. A third component of the professional mathematics teacher's knowledge base is the effective use of computer and calculator technology to promote conceptual development in the classroom. Again, many ongoing funded projects are devoted to providing opportunities for strengthening these professional knowledge bases of the teacher—but at a level of support that accomplishes only a small fraction of what is needed for excellence in our schools.

A third kind of change needed is in the kind of research performed in mathematics education. In the prevailing model, university faculty in schools of education define research problems and set out to test hypotheses; they gather data, determine significance levels for deviations from null hypotheses, and publish their findings in scholarly journals. Some effort is then made to synthesize and disseminate the research findings. We learn, from meta-analyses of hundreds of individual studies, that homework can indeed be an effective tool under some conditions for improving mathematics achievement; or that boys do somewhat better than girls on many measures of mathematical ability, but the magnitude of the difference is declining; or that access to mathematical achievement is limited for students in districts with lower socioeconomic characteristics. In paraphrasing the outcomes of traditional research studies so simplistically, I do not mean to trivialize the amount of work or expertise that has gone into them; but I do wish to emphasize that such studies do not contribute very much to solving the real problems of mathematics education in schools. They do not advance the process of change. We need, instead, research that results in working models of reform — ranging from alternative, innovative assessment methods and their consequences, to programs that benefit underrepresented populations, to successful implementation of new technologies, to conceptually based curricular innovations, to applied cognitive models. Our research should be far more schools-based, involving collaboration between teachers and university researchers, and addressing questions whose answers would remove obstacles to the reforms that are needed.

A fourth type of change pertains to the use of computer and calculator technology in the teaching of mathematics and science. It has been noted frequently that these instruments give us the opportunity to deemphasize rote processes and to emphasize meaningful, conceptual learning. I would like to stress a different point. Many people in our society have become alienated from the technology with which we are surrounded, a technology that is inaccessible and at some level frightening. This manifests itself not only in opposition to technological innovation but also in a reversion to supernatural and antiscientific belief systems, ranging from astrology and spiritualism to fundamentalist opposition to the teaching of evolution in public schools. It is difficult to understand how in this age of astronomical discovery people can really accept ancient superstitions such as astrology, until we observe that this belief fills an emotional need — a void left by people's lack of control in the technologically advanced society in which we live. Thus, it is especially important that we bring technology into the classroom in ways that give the child control over the technology, rather than the other way around. We must firmly reject models in which the student follows directions provided by the computer and is informed by the computer whether the right answer has been entered. Instead, we must enable children to use calculators and computers as tools, to solve problems they themselves define, in ways that they themselves control.

I would like to close by highlighting the urgency of the crisis in which we are working. Our educational failures have already resulted in the loss of essentially an entire generation of U.S. students to science and mathematics.

We have been fortunate that scientific research and graduate study positions in our best universities have attracted large numbers of talented people from abroad, because graduates of U.S. schools are incapable of filling these positions or not interested in doing so. Our industry has not yet collapsed from technical inadequacy, because political unrest and adverse economic conditions abroad have motivated highly trained people to come to the United States on a permanent or semipermanent basis. Although I am an internationalist, and much in favor of providing opportunities to scientists and mathematicians from abroad, I wonder how long we can escape the consequences of not developing the scientific and mathematical abilities of our own students. We have made only the slightest beginning in overcoming the underrepresentation in mathematics and science of U.S. women, blacks, people of Spanish-speaking origin, and other minority groups. This situation is significant not only for the underrepresented groups themselves, but for the whole nation as it affects our future human resource needs.

These introductory comments have touched but briefly on a number of serious difficulties in the context of mathematics education today. The chapters that follow offer additional perspectives and address some ways of meeting these challenges.

CHAPTER TWO

What Are the Issues?

ROBERT B. DAVIS   CAROLYN A. MAHER

Abstract

This book arose from a conference held at Rutgers University in November 1990, under the leadership of Dean Louise Cherry Wilkinson. Papers had been shared in advance in order to move the discussion into a more participatory, shared activity. What appears in this book is the result after *sharing and modifications had occurred.*

It was, of course, necessary to set some initial agenda for the discussion, and the paper intended to accomplish this constitutes the present chapter.

The Task of This Conference

There are plenty of recent developments in education that justify celebration, but this meeting is not a celebration. Probably the best use of the time and energy of this assembled group is to work on unfinished business. In the world of U.S. education, there is always plenty of unfinished business.

We have our own personal list of vexing problems that influence, shape, and sometimes limit our efforts to help teachers, students, and parents. We do not want to restrict attention to those issues that strike the two of us, but it may nonetheless be worthwhile to make clear the kind of problems that concern us.

A High School Class We Observed

The students were seated in the usual kind of row-and-column arrangement of desks and chairs, and were supposed to be paying attention to the teacher at the front of the room. The class was supposed to be studying percent.

In fact, probably not one student in the room had thought at all seriously about the subject. Almost no one had brought in any homework. Most students either talked with other students, or read nonclass material such as comic books, or stared into space, lost in their own thoughts. There was some effort to pretend to pay attention to the class discussion, but not much. There was no evidence that even a single student was attempting to think seriously about the subject.

The reason this class existed was, at least in part, that it was intended to prepare students for the HSPT, the High School Proficiency Test, a mandated state assessment measure in New Jersey, which a student must pass in order to graduate from high school. The textbook made no real contact with any sensible use of percent, nor was this concept really explained. It can be argued, in fact, that what the class was studying was not percent at all. For example, one question asked: "What percent is 4?" This question probably occurred to the authors as a parallel to "What percent is 1/2?" This second question is really not appropriate either, but may seem to succeed in at least eliciting the response "50%." The moment any of this is put into some kind of sensible context, the meanings change entirely—for example, 4 is clearly 50% of 8, and 1/2 is 75% of 2/3.

What bothers us about this class—besides the fact that it was frustrating for the teacher and probably harmful, not helpful, to the students—is that it probably reflects those students' attitudes toward most of school, and possibly to most of life, and that whatever was wrong was taken by the administration to be a problem that the teacher could solve by "being more effective." A few years ago there were serious efforts to understand the problem more deeply and to deal with it at a more fundamental level. For example, instead of having those students sitting in rows most of the day, one alternative was to use half of the school day for something closer to real activities—perhaps even for actually "real" activities. Some of the same students who seem to get so little out of sitting in class—and to put so little into it—behave in a far more mature and appropriate way when they are working in after-school jobs. This should send us a message.

To observe this class was disquieting, if not appalling. The textbook, the HSPT test, and even the school itself were all there to help these young people—and all too clearly they were not doing it. To treat this as a simple problem needing only more charisma, or more authority, or more entertainment from the teacher seems to misread the situation. Even if the classroom management problem were solved, one might still be left with students sitting and passively receiving; one might still be left with students making no serious effort to accomplish anything (and believing that this is sensible behavior); one might still be left with the belief that this was the most appropriate kind of learning experience for these students; one might still be left with the notion that "What percent is 4?" is a proper kind of question; one might still be left with the idea that percent was the most important thing for these students to be learning, rather than examining their own behavior and their own ways of dealing with problems. We are convinced that the changes that are needed go much deeper, and the understandings that we, as educators, require must also be at a much more fundamental level. It is this conviction that motivates our own work, and we suggest that it may not be an inappropriate focus for this conference.

Perhaps this suggestion requires some fine tuning. The reader will undoubtedly have suspected that most of the students in that class were of low socioeconomic status (SES) and probably mostly minority young people. They were, but that is not the point. Students in affluent suburbs are also asked to sit in rows and listen to the teacher explain; the mathematical content is equally pedestrian and mindless; more importantly, the kind of mental activity that is asked of the students is, in our view, equally inappropriate; and the suburban textbook may also ask: "What percent is 4?" (see, e.g., Davis, 1988). At a fundamental level, the ills of mathematics education look surprisingly similar even in very different kinds of settings.

This is not to say that the problems of minority education do not need careful attention. We agree with Kofi Lomotey when he writes:

> the status of African-American students is the most critical issue facing educators today. . . . The future of our entire country depends upon how well we educate African-Americans and other minorities in the years ahead. (Lomotey, 1990a, p. 1)

If we would modify Lomotey's remarks in any way it would be merely to emphasize that it is not only African-Americans; there are some other students in a strikingly similar plight, and these others must also be of serious concern. We fear that the general lack of success in finding more effective forms of education for minority students may lead to a situation where we come to trivialize all education as a way of making the problem less apparent — without, of course, making any progress whatsoever toward actually solving it.

We urge that the goal of this conference be to work toward that deeper kind of knowledge that may help us to improve the situation for all students, and to improve it in a fundamental way.

It may be wise to look briefly at the large educational system now in place, and how it came to take its present form.

The Background of School Mathematics

Today's school mathematics program plainly shows its historical background. Many of its features have been built upon something like the world of, say,

1900. For example, mathematical topics were often put into the curriculum in grades 1 through 6 (or possibly grades 1 through 9), because these were topics that every adult needed to know, and (in 1900, but not today) people often left school soon after completing grade 6 or so. If one were to start afresh, look at today's attendance patterns, look at the requirements on today's adults, and try to build upon today's knowledge of human learning and development, it seems absolutely certain that a very different curriculum would be put into place, and some very different kinds of learning experiences would be provided. For four decades now the United States has struggled with the task of modernizing the curricula and the pedagogical methods used in science and mathematics, especially under the leadership of the National Science Foundation and the National Academy of Science. But at the classroom level, where it would count, very little change has actually taken place. Although this conference, by itself, presumably will make no change whatsoever, it may help us all to understand a bit more deeply what are the needs and the possibilities.

If we look at mathematics education broadly, we see that what has been put in place in the United States is a very long sequence of courses, with some branching and choice at a few points. Everyone is expected to learn paper-and-pencil arithmetic; then, around grade 8 or 9, there is a branching, after which everyone in a certain large group is expected to spend a year learning "ninth-grade" algebra (which may be studied at varying grade levels in different schools); after some more branching, everyone in a certain group is expected to study trigonometry, then calculus, then (perhaps) differential equations, and so on. Not merely courses are in place; there are also rigid expectations on individual topics. Everyone (more or less) is expected to be able to solve a quadratic equation, but no one is expected to be able to solve a cubic equation. Everyone is expected to know what an acute angle is, but no one is expected to know anything about fractals. For the system, this kind of rigid and explicit specification makes for seemingly easy operation of a very large educational establishment. For students, it can make for boredom, frustration, and a sense of purposelessness that can add up to a devastatingly negative impact on their career plans and self-esteem.

What's wrong? Clearly, the kind of extreme specificity and rigid sequencing that we have described, whatever their advantages for the system, are harmful to many individual students, in part because this approach gives them the sense that they are on a virtually endless treadmill extending ahead for years and years, with no clear goal to which they can give allegiance.

But beyond that, this rigid sequence is based upon falsehood. As the studies of Lochhead and Clement on the "students-and-professors" problem, have made clear to all of us, in fact many students do not master the skills in this sequence and do not learn the key ideas (Rosnick & Clement, 1980; Lochhead, 1980). All along we have been building on a foundation that was very insecure—for many students, nonexistent. We are confronted with the mystery of how so many of these students managed to progress so far in the system without actually possessing the expected "requisite" skills.

If our remarks seem persistent, perhaps even strident, this reflects our perception that this large nationwide system of requirements, from kindergarten through college, has managed to transform the study of mathematics from what we both experienced it to be—interesting, personally rewarding, valuable, and exciting—to something that is routine, dull, unrewarding, confusing, frustrating, unchallenging despite its difficulty, and conspicuously uninspiring. Hence one part of the education of many students has lost much of its potential value and has instead become an obstacle and a source of hopelessness and frustration.[1] There is also the fact that anyone who will look at our schools, go into our subways, observe young Americans in nearly any common setting, will surely conclude that somehow our youth are not getting the start in life that our society needs them to have—not to mention how much better their own lives ought to be.

What Kind of Thinking Should Be Our Goal?

But if what is in place, from a content point of view, is highly specific and rigidly prespecified, this may not be its most limiting characteristic. That distinction probably must be awarded to the kind of mental activity that is expected of learners, and to the underlying pedagogical assumptions. One might say that for nearly all Americans *mathematics* means memorizing a specific collection of facts and algorithms, practicing them (usually in a relatively meaningless form), and, when confronted by a problem, trying to recall which of these ill-understood algorithms to use—and trying to remember how it went. We commonly find adults mumbling such things as, "Let's see—plus times plus is plus, and minus times minus is minus, but now . . . what was plus times minus?" (an actual quote from a conversation among three teachers).

To be able to use mathematics with any power, one needs a quite different collection of capabilities and habits. For example, one needs to be good at making mathematical representations of problems, either in pictures sketched on paper (Kieren, Nelson, & Smith, 1985), or in mental images, or in verbal restatements of the problem, or in modified formulations of the problem, or in actually setting out on a tabletop some physical representations, or in planning algorithms, or in writing equations, or in some other ways.

One needs to be in the habit of looking for patterns, asking onself questions about how the parts of the problem fit together (Davis, 1985), making use of a large collection of powerful heuristics, looking at alternative possible lines of attack and making shrewd comparative judgments about them, and so on. Since most of this is terra incognita to most people, it is equally unknown to most teachers and to most parents, who after all are people. Thus there is little demand for this kind of learning and very little that is done to develop it. Nor is it easy for teachers to acquire skill in developing these habits in students, when they must first try to develop them within themselves.

What Kinds of Learning Experiences?

The previous section dealt with the kind of knowledge that needs to be acquired. We turn now to the question of the kinds of experiences that are needed to help students learn what they need to know. This is perhaps the most exciting part of our own recent research. By videotaping students as they work together to try to solve mathematical problems, we have come to see how complicated the process of learning mathematics really is. A graduate student recently summed up the situation, after watching two hours of videotapes of two boys trying to learn about fractions, by saying, "I have been sitting here wondering how I ever managed to learn all of that!" Some of these complications are spelled out in detail in Maher, Davis, and Alston (1991), and further examples are given in Maher and Alston (1989), Landis and Maher (1989), and elsewhere.

Before looking at some of the complexities of "doing mathematics" and of "learning to do mathematics," we want to begin by considering a report by Dorothy Strickland and Lesley Morrow (1989) on reading and writing—specifically, on emergent literacy for young children, say at the prekindergarten or kindergarten level. Very similar issues are emerging in recent studies in reading/writing education and in mathematics education, but many people may find it easier to recognize what is involved in the case of reading and writing.

Strickland and Morrow contrast two approaches to early childhood learning, one that might well be described as "focusing on skills" and another, which Strickland and Morrow call the "emergent literacy" approach, although others might call it "holistic," "developmental," or even "problem solving." As an example of emergent-literacy learning, Strickland and Morrow cite the case of a kindergarten boy whom they call Carl. Carl wished to write a story, and wanted it to begin with something akin to "Once upon a time . . ." He was faced with the need to *write*—and therefore to spell—the word *once*. His teacher, Jan, heard him saying aloud a sequence of words, including *wishy-washy* and other words beginning with *w*. Strickland and Morrow write:

Jan . . . was delighted because she knew that Carl was drawing on his growing knowledge of the patterns in the language. Unlike the child who simply repeats *Wuh* when shown the letter *W*, Carl had learned that there are some consistencies in the language that can be relied on much of the time. *W* was [in Carl's view] at least worth exploring. . . .

Several stories later, Jan noted that the word *once* had undergone several changes. Starting with *wus*, Carl soon changed to *wuns*. When he began to feel uncomfortable with that spelling, he asked another child for help and was told to "check it out in the Cinderella book." His next attempt, *wonce*, revealed a need to hold on to the letter *w* despite what his eyes had revealed. Finally, he let go of *w* and the spelling *once* appeared. As Jan reviewed a collection of Carl's stories, she noted that at the same time that his spelling was evolving and maturing, Carl's stories were becoming more complex and more fully developed. He had learned important lessons about the consistencies and inconsistencies of his language while his ability to compose (the very reason for learning to spell) was allowed to flourish (Strickland & Morrow, 1989, p. 82).

The parallel between Strickland and Morrow's discussion of an emergent literacy approach, and the kind of mathematical experiences recommended in the National Council of Teachers of Mathematics *Standards* (NCTM, 1989) will probably have struck everyone here. For nonmathematical audiences, however, the case of reading and writing may be easier to understand. When, in mathematics, we speak of "helping the child to develop mathematical power," I fear that many readers cannot imagine what we have in mind. But look at some of the things that were involved in Carl's pursuit of *once*:

He was coming to see patterns in the language. Recognizing the appropriateness of looking for patterns is a big step forward. Careful observation of students shows all too clearly that this recognition cannot be taken for granted. (One is left to wonder if some of the "do what you're told, and do it the way we told you" orientation of much mathematics instruction may not even serve to discourage students from looking for patterns on their own initiative.)

He was learning *when to try to make use of patterns*. It is one thing to be on the lookout for patterns, and quite another to realize that looking for patterns can be a valuable tool in trying to solve some problem.

He continued to be *self-critical or self-analytical*; he was obviously monitoring his spelling of approximations to *once*, and — because he *was* watching — he came to recognize that *wus* was probably *not* a correct rendering of the sounds of *once*. He evidently thought about this enough to recognize a missing n sound, and so, on his own initiative, he changed *wus* to *wuns*. (Because of our special interest in African-American education, we wonder if the frequently harsh treatment of children's "misbehavior" may not help to cause children to become self-defensive rather than self-critical. We offer this only as a speculation, but we are desperately seeking any causes that can be identified.)

Carl was *willing to give up an established idea*—although he did not find it easy. When his eyes fell upon *once* in a book, he picked up *part* of what he was looking at but did not give up all of his earlier idea and thus produced the hybrid *wonce*. Because, however, he continued to be self-critical, *wonce* soon gave way to *once*.

One could study this episode in much greater detail and find many more things that Carl did, or needed to do, in order to move forward in his power in using the English language. Our main point is that what is involved is so complex that it would be impossible to *tell* all of this to children. Teachers must encourage it and guide it, but the acquisition of these capabilities and habits will come more from the child's experience and environment than it possibly can from explicit telling or showing.

We have a serious concern at this point. Superficially, it appears to be possible to "solve" some important educational problems by being very explicit. Educational programs and corresponding evaluation programs that are highly explicit and very sharply focused can appear to be very successful, at least in the short run. Our concern has several parts. First, we suspect that an essential aspect of learning either mathematics or reading/writing is the interrelating of the subtle processes and components we saw in Carl (and will see in further episodes from mathematical problem solving). We further suspect that this interrelating can be only learned in a context where one does it all the time (see, e.g., Miller & Gildea, 1987; Brown, Collins, & Duguid, 1989; Davis, 1989).

The problem may be even deeper. Not only may interconnections not be created, but perhaps—in a highly explicit curriculum based primarily on showing and telling and on relatively mindless practice—the more subtle among the needed capabilities and habits *may simply never be acquired at all*. In the Strickland-Morrow report, Carl had to ask a classmate and then *make a judgment about the advice he received*. Many explicit programs do not encourage this kind of judgmental listening but, instead, work toward a more passive acceptance of what a teacher says. Carl had to make an independent decision about how long, and how carefully, he would continue to monitor his current version of *once*. He was not able to merely rely on the teacher for this decision. One might say that Carl had to monitor his reading & writing performance *and even had to monitor his monitoring performance*.

A further concern is that there may be some unfortunate class differentiation involved here. The use of highly explicit school programs is probably more common in low-SES urban areas. Even if this is not true, it is almost certain that more affluent suburban children have far more opportunity to ask questions of, or to be questioned by, their parents. There is strong evidence that a significant proportion of what suburban students learn is learned at home. If, in fact, this is true, then highly explicit school programs could have the effect of raising test scores for low-SES urban children without teaching them the range of capabilities and habits that may be the most essential part of acquiring "mathematical power." If, in fact, we are teaching suburban children to think, plan, analyze, evaluate, and seek abstract patterns, and if we are teaching low-SES minority children to add and to subtract, are we not creating a truly dangerous caste system in the United States? We discuss this further in a later section.

Observing Students at Work

As mentioned earlier, our own research focuses on videotaping students at work on mathematical tasks. This takes one of three forms: (1) videotaping *task-based interviews*, where a student works on some mathematical problem, usually while one interviewer poses questions to probe more deeply into the student's thinking; (2) videotaping two or more students working together; or (3) videotaping an interaction between a teacher and one or more students. The detailed analysis of such tapes, which can take hours or even weeks (and sometimes months; see Schoenfeld, Smith & Arcavi, in press), can allow us to learn a great deal about the way students are thinking about mathematics, what their habits are, and how they deal with self-monitoring questions.

What can be learned from this kind of analysis? We present here a few examples; more detailed discussion appears in Maher, Davis, and Alston (1991).

1. *The severity of student errors.* A seventh-grade boy named Brian, in an interview with Dick Lesh, was working on a problem involving a recipe and was led to a need to compute 2 times 1/3. Brian concluded that this must be 1/6. Note that any of several glances at the meaning of these symbols would suggest that this must be false, but Brian was content with his answer. An eighth-grade girl named Lisa, in a task-based interview (also starting with a recipe problem), came to a point where she needed to add 1/2 + 1/3. Lisa concluded that 1/2 + 1/3 = 1/5. Here, too, a thought about the meanings of these symbols would have raised some questions in Lisa's mind.

One might ask, "What's new here? Didn't we already have the work of Lochhead and Clement?" We did, and the severity (and frequency) of student errors is coming to be well known, but the details are still important. We discuss this further below, in the section on methodology.

2. *How errors occur*. From these detailed observations, we also get considerable information about how errors come to be made. In Brian's interview with Dick Lesh, at one point Brian says that $1/6 + 1/3 = 1/9$. Lesh suggests that there might possibly be some doubt about this. Brian responds: "Oh, is this one of those places where you're supposed to say '$1/3 = 2/6$'?" So we see that Brian *did* possess knowledge of a correct method of dealing with this problem. What he did not possess — and this is extremely common — was enough understanding, and enough of a habit of analyzing problems, to see which different methods might be brought into play, and which would or would not be appropriate.

3. *Implications for teaching*. The preceding example, of course, raises the familiar question about the interrelations of different ideas. It also seems to us to support a teaching method long advocated by the Madison Project: *Every time* that some key decision needs to be made, the teacher should run through the reason for it. Thus, every time that one needs to add, say, $1/2 + 1/3$, the teacher would say: "In order to add these fractions, we need to arrange things so that both denominators are the same. How can we do that?" Or, if the class is faced with the equation $x + 3 = 7$, the teacher might say, "If we subtract 3 from each side of this equation, that will not change the truth set." Whenever an *action* is presented, the reason for choosing it or the legal justification for it is also presented at the same time. This approach helps to avoid the atomicity and disconnectedness that so often characterize students' ideas of mathematics.

4. *Students do think for themselves*. An extremely interesting videotape of a classroom lesson, made by Constance Kamii and her colleagues, shows some second-grade children making up their own algorithms to add and to subtract. Among the methods that they devise to solve $87 + 24$ is this one: "Eighty and 20 is 100. I knew 6 and 4 is 10, so I took 7 and 4, and that made 11, so it was 111" (made up by a second-grade girl). To solve $26 - 17$, a second-grade boy made up this method: "Twenty, take off 10, that will be 10; take away 7, that will be 3; add on the 6, that will be 9."

Similar results are reported in Cochran, Barson, and Davis (1970), in Davis (1984), and in Madell (1985). Indeed, Madell (1985), looking at all of the data, concludes that "children not only *can* but *should* create their own computational algorithms (p. 20).

5. *The complexity of what must be learned*. Consider this problem: In a box of candy, almost empty, there are 5 candies. Mary takes 2, leaving

3 for Tom. Hence Mary has taken 2 out of 5, or 2/5. But there is a second box, almost full. It contains 20 candies. Mary takes 7, leaving 13 for Tom. From this second box, Mary has taken 7 out of 20, or 7/20. What has she done altogether? She has taken 2 out of 5, or 2/5, then 7 out of 20, or 7/20. Altogether, she has taken 9 out of 25, or 9/25. So maybe you *should* combine fractions by adding the numerators and then adding the denominators? We have repeatedly seen children make up this kind of problem and solve it in this way. What do you say to the children who do this? In fact they are *not* wrong, and presumably we should acknowledge their correctness; but we may want to point out that, although inventing this way to add fractions could be useful in certain settings, there is another way that is also useful and that is the one they will more often encounter. Children can be seen making up this method in, for example, the videotape *The Polished Stones*; see Stevenson and Lee (1989).

6. *The importance of representations.* In many interviews we see clearly the important role of the various *representations* that students use. We will look in detail at some examples in Chapter 10. In particular, one can see problems or solution strategies represented in terms of verbal statements, or represented in terms of concrete models, or represented in terms of pictures, or represented in terms of mathematical notations. One can also observe how algorithms are brought to bear on these various representations. Of special importance is the preference many children show for concrete models, or for pictures. It is noteworthy how often teachers (or interviewers) reject the student's representation, even though it was appropriate and the child may have been using it successfully.

Interviews with seventh- and eighth-graders reveal a conspicuous failure of students to relate one representation to another; in many cases a student will solve a problem correctly with one representation and incorrectly when using some other representation. This suggests that we need to look more closely at what teachers are doing to help students build up a powerful capability of moving back and forth from one representation to another.

7. *Misunderstanding between people.* One very important task in dealing with mathematics is to try to build, in one's own mind, a good match for the representation that someone else is building in his or hers. This effort very often fails, resulting (among other things) in teachers rejecting many correct answers given by children (see especially Davis, Maher, & Noddings, 1990).

8. *The effectiveness of students working together.* Many tapes show two or more children working together when a class has been divided up for so-called small-group instruction. The tapes show clearly how valuable this can be and what excellent work many children do in this setting.

9. *The moods of students.* By their very nature, videotapes often reveal a great deal about the moods of students, which are often indicated by posture, inflection, how they interact with others, and what they do—or do not—undertake in the course of their work. In one instance, two fifth-grade boys, Brian and Scott, are doing excellent work in solving a problem about two pizzas (Davis & Maher, 1990). After the boys have worked for nearly 45 minutes, the teacher comes by, misunderstands what they have been doing, and gives them some incorrect suggestions on how to proceed. Their change of mood is clearly apparent on the videotape: From being persistent, resourceful, creative, and interested, they now divorce themselves from serious thinking about the task at hand and become uninvolved. Without the videotape, all of this would have gone unobserved (Maher & Davis, 1990).

10. *The Law of the Ineffective Middle.* Often, one can build a valuable discussion on concrete and familiar ground, or else on the foundation of precisely stated abstract terms. Children can and do deal well with either of these approaches. Our observations, however, show that teachers and interviewers all too often try for a kind of middle ground that is rarely effective. It is not concrete enough (or, really, not *familiar* enough) to count as "concrete/familiar," nor is it precise enough to count as "precise/abstract." As one example, Brian, a seventh-grade boy, is dealing with a problem in fractions where the unit he is using is really two yellow hexagons (in the set of Pattern Blocks). Apparently this unit is *not* sufficiently familiar to him for him to feel truly at home with it, and it gives him a great deal of trouble (Maher, Davis, & Alston, 1991). Now one *could* move toward a precise/abstract redefinition of the problem, establishing very clearly the notion of a *unit* and then talking in terms of this unit. Alternatively, one could move to a more familiar concrete idea, which, in fact, is what the interviewer actually does; changing language, she asks her questions in terms of chocolate bars—with which, of course, the boy is more familiar.[2]

11. *Failure to establish the basic task.* Studying the videotapes reveals that teachers and interviewers often fail to establish the basic task on which students are asked to work. In one interview, a student is asked to mark fractions on a number line "in an orderly way" (Maher, Davis & Alston, 1991). No attribute of the order is specified. What the student does is to use his awareness that the fraction k/a—where $k > 0$—is smaller than k/b if and only if $a > b$. Based on this, he marks k/a on the number line in the interval between the integer $k - 1$ and the integer k, then uses the inequality just mentioned to get a correct ordering between $k/2$, $k/3$, $k/4$, and so on.

In fact, the student has carried out the prescribed task in a perfectly sensible way. Imagine how one delivers mail to east-west streets in Manhattan; first one finds the correct street number (West 54th Street, for example), then one finds the correct street address (523 West 54th Street, say), and then, within that building, one finds the correct apartment number. Of course, what had been intended had been an ordering based on size, and such that lengths could be used for addition; thus, 1/3 + 1/3 should, if the lengths were added, give a correct indication of where 2/3 would be marked. Similarly, 1/3 + 1/3 + 1/3 should, if lengths are added, come out to the integer point 1. But this basic requirement was never stated to the student. This is a very typical situation.[3]

12. *Mapping input data into schemas.* The broad outline of our theory of human information processing is essentially the framework presented in Davis (1984). The main points are the creation of representations in one's mind (sometimes with the help of paper, Pattern Blocks, etc.); the mapping of input data into these representations; judging the adequacy of these representations and these mappings; and the fact that representations are built up largely using simple and very familiar component parts. Tape recordings of students show clearly that much of the difficulty that people say they find in "arithmetic" can be located more precisely in the specific step of trying to map reality into the "variables" or "slots" in representations. As one example, a problem speaks of "a 3-mile race," "Mary running 3/4 of a mile in the first 5 minutes," and "Mary running 2/3 of a mile in the next 5 minutes." A student draws a line segment to represent the race. Then, forgetting what the line was intended to represent, he interprets the first statement by marking Mary's position after 5 minutes at a point 3/4 of the way along the line segment, which *ought* to mean "3/4 of the *race*," not "3/4 of a mile." The taped records are replete with instances of errors at this precise point of mapping input data into appropriate slots in representations.

13. *Noun knowledge versus verb knowledge.* The interrelations between our underlying theory and the details of the taped interviews are obviously of great interest to us. We mention one other; we have talked in the past about the distinction between *verb* knowledge and *noun* knowledge (Davis, 1984, p. 36). Commonly, one learns to *do* something as a sequence of actions, without being able to step back and look at what one has done, to see it as a kind of "thing." We have described this by saying that students (or anyone, for that matter) will often acquire "verb" knowledge before they develop "noun" knowledge.

The tapes show some stunning examples. In one tape (Martino, in process), four second-graders are asked to find how many unit cubes would be required to construct a 4-by-4-by-4 cube (actually using Dienes base 4 MAB blocks). The children put a "flat" (a 4-by-4-by-1 array) down on a piece of paper, trace around it, and fill it in with unit cubes — this despite the fact that the flat is ruled so as to show how it might be constructed from unit cubes, and despite the fact that units, "longs" (4-by-1-by-1 arrays), and flats are right there in front of their eyes. They seem unable, or at least unwilling, to make use of aids that they have not painstakingly constructed themselves. After they have built one flat themselves out of units, they begin to make use of longs and flats to finish the job—but they do not use them at the outset.

14. *In class, things happen fast.* In analyzing tapes, we have been taken by surprise by the rapid flow of actions and the dense stream of information. Many things happen almost simultaneously, and the sequence is extremely fast. A child trying to find a Cuisenaire rod one-third as long as a blue rod may start to reach for a purple rod; then the child's hand changes direction and picks up a light green rod. A child may say two words, then be interrupted by another child. A child may glance at a fraction written on the board and apparently get guidance from that glance to assist in writing some other fraction on the paper. A child may write an answer, then immediately erase it and write something different. The original answer may have been on view for two or three seconds. If the teacher did not see it during those two or three seconds, then the teacher never got to see it at all. Careful analysis of a tape can reveal an information stream so dense that no teacher—not even one with eyes in the back of his or her head—could have taken it in during the lesson.

The pace of this information can be important. A few years ago, in working with a computer that attempted to give children guidance by using audio remarks from a floppy disk, we ran into the following problem: The computer sound system would say "A" and ask the child to point to the letter A on the computer screen, which had a touch panel. The child would point to some other letter—say, D. The computer would read a message from the audio disk, but the child would be faster and would move his finger to point to the A—just in time to hear the message, "No, that's a D. Please point to the A." Thus, because of the slower pace of moving the read head to the proper location on the audio disk, the child would be pointing at an A while hearing a message asserting it that was *not* the letter A. This reading program was in fact found to impede a child's learning to read (one of the rare instances of an innovation that actually proved harmful!). There were undoubtedly many reasons for this failure, but the slow response time of the audio signals probably contributed. It was another instance of the very fast pace of things happening in a classroom. By ignoring all of this, one loses an opportunity to track far more closely the child's patterns of thinking and problem solving. In one taped interview, when asked to see which was larger, 1/3 or 1/4, Brian *immediately* put down a train of Cuisenaire rods consisting of one orange rod followed by one red rod (Maher, Davis, & Alston, 1991). This is in fact the shortest (and in that sense the most efficient) train that can be used to solve this problem. Brian put it in place without a moment's hesitation. How did he know to make this choice? By not hesitating or making trial attempts, Brian was, through this behavior, displaying some significant knowledge; it took him less than a second to do it.

Programs for Urban Minority Students

As indicated earlier, we agree with Lomotey that the status of minority students is a matter of paramount concern. There is no question that there is a problem here, and a very important one. To cite some of Lomotey's description of the situation:

> The underachievement of African-American students in public schools has been persistent, pervasive, and disproportionate. Since the introduction of standardized tests as measures of achievement, African-American students have always lagged behind their white peers. On average, by the sixth grade, African-American students trail their white peers by more than two years in reading, math, and writing skills as measured by standardized achievement tests.
>
> If we look at statistics regarding African-American success in high school, a similarly disturbing pattern emerges. . . . In New York City, the dropout rate for African-American males is more than 70 percent.

Lomotey writes both for the purpose of understanding the problem and for the purpose of improving the situation. Those are our purposes also. But a very serious danger lurks beneath any honest attempt in this direction. Almost any sentence that one can write may contribute, not to a deeper understanding, but to greater confusion, to *mis*understanding, and to inappropriate responses. Is the issue nowadays so emotionally charged that one cannot talk about it at all? If so, then the application of our single most valuable tool, human intelligence, cannot be brought to bear on a problem that all of us want desperately to solve.

If we were politicians running for reelection, we probably would not dare to quote Lomotey. Indeed, if he were a politician running for reelection, he might not have dared to write those words in the first place. Imagine the negative thirty-second commercial that an opponent could make from a sound bite such as those we have just set down on paper!

Even the combination of words that one uses can lead us astray—for example, the designation *African-American*, which seems to make important reference to something *African*. Could anyone listen to Nelson Mandela and conclude that there is a mysterious "African" essence that imposes some sort of limitation on a command of the English language?

Of course one can respond that *African-American* does *not* say *African*. It is not the experience of people in Africa that is involved but the experience of Africans who live in the United States. Well, Barbara Jordan is a person of African ancestry who lives in the United States. How would you describe her educational success? Or Billy Taylor's? Or James Baldwin's? Or Bill Cosby's?

To solve the problem here, we must first try to understand it, and that means we must find ways to talk about it. It is clear that black children can learn. It is clear that, at present, many of them are not learning many of the things they need to learn—and which, we are convinced, they would *want* to learn if we could find the right kinds of experiences to provide for them.

In the case of mathematics, especially, there is a specific danger: Some of the proposals that are intended to improve the situation may in fact have the potential of making things worse. If black inner-city children are restricted to drilling on arithmetic, while white and Asian children in the suburbs are playing with computers and learning an awe-inspiring creativity, are we building a future equality of opportunity? And if those suburban children are facing similar restrictions on the content that they learn in school, and get (as many of them apparently do) the "serious" part (which is, of course, the playful and creative part) of their education *outside* of school, where does that leave us? Confused, at the very least, and in need of a major improvement in our understanding of the situation.

In point of fact we do not like either label, "African-American" or "black"; we are talking about young people who do not have much success in school, and we are trying to find those patterns in their lives—in school, at home, on the street, in stores, watching TV, or wherever—that are helpful and those patterns that are harmful. Two questions need to be answered: (1) Why don't these students have more success in school mathematics: (2) Are we implementing the kinds of programs that will make things better?

Concerning the first question we have very few data, but—having thought about the matter during years of working with urban minority students—we do have some conjectures. One or more of the following may be at work:

1. Perhaps the students' attitude toward work itself is part of the problem, and this may be a reflection of parental attitudes. At University High School ("Uni," for short), where one of us taught for some years, most students were the children of judges, lawyers, concert pianists, professional research mathematicians, physicists, and so on. Their parents in large part defined themselves by their work—as, say, Beethoven or Picasso must have done. The students seemed to have picked up this attitude and saw themselves as future doctors, physicists, computer scientists, and so on. With this orientation, working hard to get the deepest possible understanding of mathematics made perfectly good sense, and that is what the students mostly did. (One is not surprised that Nat King Cole's daughter became a singer.)

By contrast, the parents of urban minority children will, in many cases, probably have had different experiences in the world of work. As a result, one would expect that many of them have come to a different attitude and, in many cases, do not define themselves by the work that they do. They may well — and for good reason — view work as an unpleasant evil that is necessary in order to earn money. This attitude, too, is probably passed on to their children. In fact, we have encountered many urban students who seem to do only enough work to satisfy the teacher and to do none at all when the teacher's authority does not require working. This is not defined by race; some of the most conspicuous instances have been white children, non-Hispanic. By contrast with this attitude of working only when supervised, Uni students are usually eager to spend the odd free moment working on something like mathematics, because it is *their* goal to achieve more skill or deeper understanding. They are not doing it because of the presence of a supervisor. (In fact, this is a major distinction, affecting not only the amount of work and learning that take place but also the nature and quality of that work; we do things one way when we want to get something done and over with, and do things quite differently when we want to get everything *right* and to feel proud of the result.)

2. Distinct from one's attitude to work in general, there is also the question of *your own view of your future self*. This may be part of what Booker Peek calls "political education." Peek describes it this way: "Political education . . . is a broad term I am using to apply to the culture, history, desire, and aspirations of a child that have been shaped in great part by the parents, the grandparents, all of his or her environment" (Lomotey, 1990b, p. 14). The question can then be cast in the following terms: Something definite is offered by the school. For the calculus course at Uni, this included considerable analysis of what might be called the philosophical questions of calculus — for example, how can the theory of limits allow an absolutely precise definition of the limit of a sequence, if the terms of the sequence may only give you some kind of approximation? (The answer depends on using the Law of Trichotomy and the method of Indirect Proofs.) For many Uni students, with their views of their future selves, this was a welcome kind of knowledge. But for many urban minority students it surely would not be. They would find it impossible to fit this kind of knowledge into their own view of their own future selves.

3. It is difficult, if not impossible, to acquire much power in mathematics if one is not both skeptical and inquisitive. Some homes encourage this attitude. One family asked their son almost daily, "Did you ask a good question in school today?" Other homes discourage this orientation and argue instead that you should just do what you're told, even if you don't understand it.

4. Another possible cultural difference may lie in the area of self-criticism. It is hard to learn mathematics unless one analyzes one's mistakes. But this means that one must seek out one's mistakes and try to understand them. This is quite different from an attitude of defensiveness, in which one tries to deny, forget, or conceal one's errors. Many urban environments are harsh and punitive, and this atmosphere hardly encourages acknowledgment of one's errors. The resulting attitude may make learning mathematics considerably more difficult.

5. A variant of the preceding mechanism has been identified by Treisman; whereas the Oriental students whom he observed were quick to compare notes with other students for their mutual advantage, African-American students were very reluctant to do so and tended to assume a kind of John Wayne attitude of self-sufficiency. Among the students whom Treisman studied, cooperation worked much better.

On the other question, of whether we are implementing the most appropriate school programs to help minority students, we have indicated earlier our concern that highly explicit, sharply focused programs, though tending in the short run to raise test scores, may actually be harming students in the longer run by (among other things) encouraging an attitude of passivity and acceptance instead of curiosity and skepticism. The kind of direct observational studies that we are now carrying out may come to shed some important light on this question.

Research Methodology

The kind of research in which we are engaged has some aspects that deserve explicit mention. First, the *kind* of *knowledge* that is developed has often been overlooked. We are all familiar with the kind of knowledge that can be expressed as generalizations, usually cast in the form of natural-language statements or mathematical formulae. Examples include $F = ma$, "Chlorides tend to be soluble, sulfides tend to be insoluble,", and "The planets travel on elliptical orbits around the sun." We all know, but do not often think about, the fact that there is another, quite different, kind of knowledge. What is a "force"? What is "mass"? What is acceleration"? What is a "planet"? What is an "orbit"? Those of us who feel familiar with these ideas may see nothing about them that could possibly be problematic. Because we are comfortable with the basic things involved, we can count them, measure them, and so on, doing with them whatsoever we may consider worthwhile.

If one looks at history, these "things" do not start out as well defined at all, nor are their identities all that simple even today. What is true for history is also true for individuals. The process of increasing human knowledge is not merely a matter of measuring, counting, and generalizing. An even more basic process is that of increasing people's basic collection of "known things" or "basic ideas" or "assimilation paradigms" or "schemas" or—as George Lakoff calls them—"basic metaphors." One looked into the evening sky and saw small points of light. But these "things" are not to be thought of as "small points of light." At some point many of them became "stars," such as "the Evening Star"—which, of course, subsequently ceased to be a star and became a "planet." Gaining relatively direct knowledge of these "basic things" is perhaps the really fundamental level of research. A few years ago this kind of activity tended to be denigrated as "merely anecdotal" or (in the case of anthropology) as "mere travelogues."

There is a basic point here of considerable importance. We would argue (and have argued; see, e.g., Davis, 1967, 1984) that there are at least two information-processing activities that humans employ in thinking. First, there is the process that makes use of words or other shared symbolisms; but there is also a second and quite different process that makes little or no use of publicly shared symbolisms—namely, the process of mapping input data into our own personal collection of assimilation paradigms (to use Piagetian language) or "metaphors" (to use the language of Lakoff and Johnson). These processes operate quite differently. In the mapping process we are often not consciously aware of why we have a particular opinion or make a particular choice. It just "seems right." We have considered some situation; aspects of it have been mapped into some assimilation paradigm. We therefore see it as "just like" or "similar to" some recalled earlier experience (probably not some single experience but, rather, some accumulated collection of earlier experiences that have given rise to the more abstract, generally representative assimilation paradigm). We are usually unaware of this process, but it may—on, say, the occasion of our meeting a new person—cause us an uncomfortable sense of concern, fear, or dislike. Polya proposed using this process in a more explicit form, in order to improve one's ability to solve novel problems, by asking oneself, when confronted by an unfamiliar problem: "What does it remind me of? What have I ever seen that might be somewhat similar to this new problem?" (Incidentally, one could argue that one of the central tasks of Freudian psychotherapy deals with the relation between these two kinds of knowledge, that which is verbal or rule-governed, and that which is embodied in a nonverbal way within one's personal collection of assimilation paradigms, created as "summaries" of earlier experiences.)

The kind of knowledge represented by one's personal collection of assimilation paradigms, and one's use of this mapping process, is very different from the kind of knowledge represented by generalizations. We would argue that, in most cases ($F = ma$ and $e = mc^2$ being conspicuous exceptions), the knowledge that is, or can be, stored in the form of generalizations is relatively superficial compared to the nonverbal knowledge that is embodied in one's personal collection of assimilation paradigms. Everyone knows many examples. Here is one: Public television recently broadcast an eleven-hour presentation entitled "The Civil War." One historian remarked that "to understand America, you have to understand the Civil War." But what is the nature of that understanding? Not something that one could easily put into words. That the war was "serious"? That after the war more than half of the budget of one state was devoted to paying the cost of artificial limbs for veterans who had lost one or more arms or legs? That some parts of the North also tried to secede? That Lincoln seemed to face defeat for reelection? That some Northern generals seemed unwilling to order their soldiers to attack (despite Lincoln's ordering them to do so), thereby prolonging the war for months, if not years? The kinds of letters that soldiers wrote home to loved ones? The number of soldiers who died from disease? The creation of entirely new forms of warfare, including iron-clad ships? What kind of "understanding" are we talking about here, anyhow? Yet after watching the eleven hours (which a record number of Americans actually did), it was indeed fair to say that one did have a better understanding of America, and that it was, indeed, precisely because one had a better understanding of the Civil War. What one had was not some sort of generalization but, rather, a very much richer collection of assimilation paradigms, many of them extremely meaningful and unforgettable.

It took public television eleven hours to present this story precisely because there was no way that it could be encapsulated in a modest number of words or symbols. We would argue that the knowledge that teachers require is more often of this sort. It is only rarely stateable in terms of generalizations. It is a kind of knowledge closer to what a historian knows than to the contents of a textbook on physics.

We said "the contents of a textbook on physics" rather than "what a physicist knows" because we strongly suspect that a very large part of the knowledge of the best physicists consists precisely of their extensive collection of assimilation paradigms, of which they are often unaware but which they use in frequent and essential ways. Try asking a physicist why water spirals out the drain of a bathtub, and pay special attention to how he or she explains an "inertial frame of reference."

Both "knowledge represented by generalizations" and "knowledge represented by a collection of assimilation paradigms" play important roles. The history of most sciences shows some oscillating between alternative norms, alternative expectations, or alternative imperatives; toward the end of the nineteenth century, physicists frequently expressed the opinion that the key *phenomena* were now known, the key entities or concepts were established—which is to say that the necessary collections of assimilation paradigms were in place—and all that remained was the more precise measurement of key physical constants. Within two decades, most of this certainty had been swept aside by uncertainties and contradictions that led to relativity and to quantum mechanics. A redetermination of the basic "things" was once again the main order of the day. Physicists had to work hard to enlarge their personal collections of assimilation paradigms, a task that did not turn out to be easy.

The use of videotapes of human mathematical activity shows us a great diversity of "things that go on" that do not at all match up with the entities usually postulated in order to describe human thinking. We find ourselves thinking of our work rather in the fashion of the explorers of earlier times, who traversed unfamiliar geographic domains and came back with new knowledge of rivers, mountains, and strange new plants and animals. This aspect of our work is discussed in more detail in Davis (1990); see also Lakoff and Johnson (1980), Lakoff (1986), and Johnson (1987).

A second aspect of this kind of work is also important and is often misunderstood. When one goes into a school to videotape students doing mathematics, and then analyzes these tapes in careful detail, one is simultaneously engaged in research, data collection, data analysis, theory building, curriculum design (since you will surely try to find ways to avoid or remediate any observed weaknesses and to take advantage of any observed possibilities), pedagogical innovation (for the same reasons), and teacher education. No one of these can be (or should be) separated off from the others. The basic questions you are addressing are: How do students think about mathematics and how can we make this process work better? All the components listed are necessarily a part of this process. Indeed, there are really no defendable boundaries between the various parts.

Finally, in this kind of research there is no way to know in advance what you will find. Virtually everyone who has studied the tapes has been surprised by what they have seen. You may think you know what happens when students think about mathematics—but look carefully and you will always find things that are both new and unexpected.

We could summarize our methodological concerns by saying three things: First, a different kind of knowledge is involved, not generalizations but, rather, an increasing of one's collection of assimilation paradigms. Second, all of this work constitutes a seamless whole; there is no separation between research, teacher education, and curriculum and pedagogical innovation. Finally, one truly does not, and cannot, know in advance what one will find. Our videotaping of students working on mathematics is an exploration into unknown territory. But there is one certainty: Important things are there, waiting to be found, and they are often useful in the task of improving the teaching and learning of mathematics.

Notes

1. There is evidence that the situation has in fact grown worse. Consider this description, from Shujaa (1990):

> for much of the education reform that occurred throughout the 1980's . . . [p]olicies that increased student accountability were among the most popular reforms. Student accountability policies are intended to improve educational outcomes by imposing more stringent standards for earning course credit, grade-to-grade promotion, and graduation. High school graduation requirements, for example, were increased in 45 states. Other types of policies that increased standards for students were new or modified high school exit exams, promotion or gate tests, and minimum grade point average. (Shujaa, 1990, p. 85; Fuhrman, Clune, & Elmore, 1988; see also Bliss & Carrasco, 1990)

How did all of this work out in practice? Shujaa reports:

> Increased course requirements and exit exams often produced lowered teacher expectations for students who were considered to be "non-academic." Teachers who felt they did not have the necessary resources and support to meet the existing standards tended to view policies that increased student accountability with cynicism and frustration. "Policies that mandated student testing were often perceived negatively by teachers who were forced to devote instructional time to preparing their students to take tests. These practices often conflicted with teachers' ideas about what should be emphasized in the classroom" (Shujaa, 1990, p. 101).

In short, what should be one of the most important resources in the classroom—the knowledge and judgment of the teacher—was rendered inoperative. Teachers have far less decision-making freedom today than they had twenty or thirty years ago. Even more forceful and more pointed criticism of this development has been presented by Daniel Koretz (1988). So much is *required* and *covered on tests* that little freedom to pursue anything interesting remains.

These rigid requirements and specifications of recent years contrast sharply with the humanistic concerns of a few years earlier, where students seemed to be perceived in a more holistic way, and with a recognition that the ability to do long division might not be nearly as important to many children as the feeling that the teacher cared about them and wanted to help them, that they themselves had some reasonable measure of control over their own lives, and that there was some promising vision of their personal future that depended upon some measure of effective learning. See, for example, Kelley (1951), Mearns (1958), Goodman (1960), Boulle (1960), Cantor (1961), Goodlad and Anderson (1963), Henry (1963), Holt (1964), Hentoff (1967), Kozol (1967), Schools Council (1969), Dennison (1969), Glasser (1969), Postman and Weingartner (1969), Herndon (1968), Hertzberg and Stone (1971), Featherstone (1971), Richmond (1973), and Ridgway and Lawton (1973).

What happened to the humanistic, or holistic, approach of the 1960s? Little effort seems to have gone into answering this question, but it is an important one, and deserves better. Will we ever find ways to learn from the history of what happens in schools? Or are community pressures and internal conflicts so great, and so situation-specific, that nothing general *can* be learned? Looking at an individual school it sometimes seems so. Nonetheless, when a large nationwide concern for the whole life and goals and self-concept of each individual student exists, flourishes, and then disappears, it surely seems as if something systematic is at work somewhere.

2. These observations, together with a recent analysis by Julian Weissglass (1991), may have helped us to understand the work of Mellin-Olsen, which had confused us for some time. Mellin-Olsen distinguishes what he calls capital-A "Activities" from lower-case-a "activities." A learning experience is a capital-A Activity if it is experienced by the student: (1) both in relation to the individual history of the student and also to the history of the student's culture; (2) in such a way that skill learning is part of some larger project that is of interest to the student; and (3) within some context of social cooperation, so that both the individual student and also the group with which he is working are the gainers.

Mellin-Olsen seems to be discussing some important matters, but in a way that we had not felt we understood. Somewhat similar references to "culture" occur in the analyses of several other scholars (for example, D'Ambrosio), and have always confused us, in part because most of the successes in school mathematics with which we are familiar have, at least at first glance, seemed to be remarkably culture-free. But after studying Weissglass's valuable analysis of Mellin-Olsen's recent book (Mellin-Olsen, 1987) and looking carefully at some of these videotapes, we have a possible translation of Mellin-Olsen's central idea. For Brian, work with the Pattern Blocks may have been a lower-case-a activity but not a capital-A Activity because yellow hexagons were nowhere near as familiar to him as were chocolate bars — and, in particular, the idea of dividing up pairs of yellow hexagons was not a really familiar task, whereas the idea of dividing up a chocolate bar definitely was. It was easy for Brian to overlook the fact that his unit was a *pair* of yellow hexagons; he would not have been likely to switch and decide that his unit was *one-half* of a chocolate bar. The issue may not be merely the singleness of the chocolate bar; would Brian have divided up a *family* of mother, father, son, and daughter, and inadvertently started treating *half* a family as his unit?

3. In fact, it is part of a common teaching strategy. Where we would argue that good teaching should be clear on the basic task but should leave it to the student to invent appropriate methods of solution, it is more common to find teachers leaving the task ill-defined but showing enough method that the student is able to carry out the task by imitation. No wonder, then, that students cannot tell which method they should employ in which situation. Their in-class decision making has been done by imitating what they have seen the teacher do. They have had little or no in-class experience with analyzing situations and then using their own analyses as a basis for devising a strategy of solution. (Of course, most teachers themselves have had little or no experience with this, either.)

References

Bliss, J. R., & Carrasco, M. C. (1990). Changing an urban school: Problems of capacity and power. In K. Lomotey (Ed.), *Going to school: The African-American experience*. Albany, NY: SUNY Press.

Bliss, J. R. (1990). Strategic and holistic images of effective schools. In J. R. Bliss, W. A. Firestone, & C. E. Richards (Eds.), *Rethinking effective schools: Research and practice* (pp. 43-57). Englewood Cliffs, NJ: Prentice-Hall.

Boulle, P. (1960). *The test*. New York: Vanguard.

Brown, J. S., Collins, A., & Duguid, P. (1989). Situated cognition and the culture of learning. *Educational Researcher, 18*(1), 32-42.

Cantor, N. (1961). *The dynamics of learning.* East Aurora, NY: Henry Stewart.
Cochran, B. S., Barson, A., & Davis, R. B. (1970). Child-created mathematics. *The Arithmetic Teacher, 17* (March), 211-215.
Coleman, J. S., Campbell, E., Hobson, C., McPartland, J., Mood, A., Weinfeld, F., & York, R. (1966). *Equality of educational opportunity.* Washington, DC: U.S. Government Printing Office.
Comer, J. P. (1980). *School power.* New York: Free Press.
Davis, R. B. (1967). Mathematics teaching, with special reference to epistemological problems. Monograph 1, *Journal of Research and Development in Education.*
Davis, R. B. (1984). *Learning mathematics: The cognitive science approach to mathematics education.* London: Croom Helm.
Davis, R. B. (1985). A study of the process of making proofs. *Journal of Mathematical Behavior, 4*(1), 37-43.
Davis, R. B. (1988). Is percent a number? *Journal of Mathematical Behavior, 7*(3), 299-302.
Davis, R. B. (1989). The culture of mathematics and the culture of schools. *Journal of Mathematical Behavior, 8*(2), 143-160.
Davis, R. B. (1990). The knowledge of cats: Epistemological foundations of mathematics education. *Proceedings of the Fourteenth PME Conference.* (Vol. 1, pp. 1-24). Mexico City: International Group for the Psychology of Mathematics Education.
Davis, R. B., & Maher, C. A. (1990). What do we do when we "do mathematics"? In R. B. Davis, C. A. Maher, & N. Noddings (Eds.), *Constructivist views on the teaching and learning of mathematics.* Monograph 4, *Journal for Research in Mathematics Education.* Reston, VA: National Council of Teachers of Mathematics.
Davis, R. B., Maher, C. A., & Noddings, N. (1990). *Constructivist views on the teaching and learning of mathematics.* Reston, VA: National Council of Teachers of Mathematics.
Dennison, G. (1969). *The lives of children.* New York: Random House.
Featherstone, J. (1971). *Schools where children learn.* New York: Liveright.
Fuhrman, S., Clune, W., & Elmore, R. (1988). Research on education reform: Lessons on the implementation of policy. *Teachers College Record, 90*(2), 237-257.
Glasser, W. (1969). *Schools without failure.* New York: Harper & Row.
Goodlad, J. I., & Anderson, R. H. (1963). *The nongraded elementary school.* New York: Harcourt, Brace, & World.
Goodman, P. (1960). *Growing up absurd.* New York: Vintage.
Gordon, E. (in press). *Defiers of negative prediction.* New Haven: Yale University Press.
Haynes, N. M., & Comer, J. P. (1990). Helping black children succeed: The significance of some social factors. In K. Lomotey, (Ed.), *Going to school. The African-American experience.* Albany, NY: SUNY Press.
Henry, J. (1963). American schoolrooms: Learning the nightmare. *Columbia University Forum, 6*(2), 23-30.
Hentoff, N. (1967). *Our children are dying.* New York: Viking.
Herndon, J. (1968). *The way it spozed to be.* New York: Simon and Schuster.
Hertzberg, A., & Stone, E. F. (1971). *Schools are for children.* New York: Schocken.
Holt, J. (1964). *How children fail.* New York: Pitman.
Johnson, M. (1987). *The body in the mind.* Chicago: University of Chicago Press.
Kamii, C. (undated). *Double column addition: A teacher uses Piaget's theory.* Birmingham, AL: Promethean Films South. (VHS videotape).
Kelley, E. C. (1951). *The workshop way of learning.* New York: Harper.

Kieren, T., Nelson, C., & Smith, G. (1985). Graphical algorithms in partitioning tasks. *Journal of Mathematical Behavior*, *4*(1), 25-36.

Koretz, D. (1988). Arriving in Lake Wobegon: Are standardized tests exaggerating achievement and distorting instruction? *American Educator*, *12*(2), 8-15, 46-52.

Kozol, J. (1967). *Death at an early age: The destruction of the hearts and minds of Negro children in the Boston public schools*. Boston: Houghton Mifflin.

Lakoff, G., & Johnson, M. (1980). *Metaphors we live by*. Chicago: University of Chicago Press.

Lakoff, G. (1986). *Women, fire, and dangerous things*. Chicago: University of Chicago Press.

Landis, J. H., & Maher, C. A. (1989). Observations of Carrie, a fourth-grade student, doing mathematics. *Journal of Mathematical Behavior*, *8*(1), 3-12.

Lochhead, J. (1980). Faculty interpretations of simple algebraic statements: The professor's side of the equation. *Journal of Mathematical Behavior*, *3*(1), 29-38.

Lomotey, K. (1990a). (Ed.). *Going to school. The African-American experience*. Albany, NY: SUNY Press.

Lomotey, K. (1990b). An interview with Booker Peek. In K. Lomotey, (Ed.), *Going to school: The African-American experience*. Albany, NY: SUNY Press.

Madell, R. (1985). Children's natural processes. *Arithmetic Teacher*, March, 20-22.

Maher, C. A., & Alston, A. (1989). Is meaning connected to symbols? An interview with Ling Chen. *Journal of Mathematical Behavior*, *8*(3), 241-248.

Maher, C. A., & Davis, R. B. (1990). Building representations of children's meaning. In R. B. Davis, C. A. Maher, & N. Noddings (Eds.), *Constructivist views on the teaching and learning of mathematics*, Monograph 4, *Journal for Research in Mathematics Education*. Reston, VA: National Council of Teachers of Mathematics.

Maher, C. A., Davis, R. B., & Alston, A. (1991). Brian's representation and development of mathematical knowledge: A four year study. *Journal of Mathematical Behavior*, *10*(2).

Mathematical Sciences Education Board. (1990). *Reshaping school mathematics*. Washington, DC: National Academy Press.

Mearns, H. (1929; reprinted 1958). *Creative power: The education of youth in the creative arts*. New York: Dover.

Mellin-Olsen, S. (1987). *The politics of mathematics education*. Dordrecht: D. Reidel.

Miller, G. A., & Gildea, P. M. (1987). How children learn words. *Scientific American*, *257*(3), 94-99.

National Council of Teachers of Mathematics. (1989). *Curriculum and evaluation standards for school mathematics*. Reston, VA: Author.

National Research Council. (1989). *Everybody counts: A report to the nation on the future of mathematics education*. Washington, DC: National Academy Press.

Postman, N., & Weingartner, C. (1969). *Teaching as a subversive activity*. New York: Delacorte.

Richmond, G. (1973). *The micro-society school: A real world in miniature*. New York: Harper & Row.

Ridgway, L., & Lawton, I. (1973). *Family grouping in the primary school*. New York: Ballantine.

Rosnick, P., & Clement, J. (1980). Learning without understanding: The effect of tutoring strategies on algebra misconceptions. *Journal of Mathematical Behavior*, *3*(1), 3-28.

Schoenfeld, A. H., Smith, J. P., & Arcavi, A. (in press). *Learning: The microgenetic analysis of one student's evolving understanding of a complex subject matter*. Manuscript submitted for publication.

Schools Council. (1969). *Mathematics in primary schools*. London: Her Majesty's Stationery Office.
Shujaa, M. J. (1990). Policy failure in urban schools: How teachers respond to increased accountability for students. In K. Lomotey (Ed.), *Going to school. The African-American experience*. Albany, NY: SUNY Press.
Stevenson, H. W., & Lee, S. (1989). *The polished stones*. Ann Arbor: University of Michigan. (VHS videotape).
Strickland, D. S., & Morrow, L. M. (1989). Developing skills: An emergent literacy perspective. *The Reading Teacher*, October, 82–83.
Weissglass, J. (1991). Reaching students who reject school: A need for strategy. (Review and analysis of Mellin-Olsen, *The politics of mathematics education. Journal of Mathematical Behavior*, *10*(3), 279–297.

CHAPTER THREE

Constructivism and Caring

NEL NODDINGS

Abstract

Constructivists believe that knowers must construct their own knowledge, but the knowledge produced by a given set of constructions may not be as adequate, accurate, or powerful as that produced by further construction. For that reason, as the early pragmatists recognized, constructivism requires a commitment to continued inquiry. This commitment is essentially ethical because it requires us to decide what the purpose of our investigation is and to tailor it to fit our purpose. When we work in this way, we must be aware of the limits on our claims and convey them honestly to colleagues. Constructivist pedagogy requires a similar ethical commitment. Understanding and accepting student purposes, we ask different questions of different students and urge them to design their investigations so that they are adequate for their own well-considered purposes. Constructivist pedagogy requires the receptivity and responsiveness characteristic of an ethic of care. When we care for our students, we take an interest in their purposes as well as our own. Care thus leads us to connect school mathematics not only to "real" mathematics but to real life in all its breadth and wonder. Teachers who would care in this way obviously need time to develop relations of trust with their students, and they need time to converse — to share and to listen. In an atmosphere guided by constructivism and care, much more is learned by both students and teachers than can be prescribed at the outset. Most important is that both become competent, caring, loving, and lovable people.

There are two ways in which constructivism and other current cognitive views require an ethic. First, as C. S. Peirce (see Thompson, 1963) recognized, any epistemological position that defines truth as the outcome of continued inquiry requires the commitment of inquirers to continue investigation so long as legitimate objections can be raised to their conjectures. Second, when constructivism is interpreted pedagogically, it creates a need to consider the purposes and needs of students. Constructivists recognize that students must construct their own mathematical knowledge, but why should students engage in mathematical activity at all? What ethical obligations do teachers have to meet the nonmathematical needs of students, and how far should teachers push students to continued inquiry?

In this chapter, I will discuss the first issue briefly and then consider in some depth caring as an appropriate ethical framework for constructivist pedagogy.

Knowledge and Truth as Outcomes of Inquiry

Pragmatic philosophers and like-minded thinkers (many cognitivists) regard both knowledge and truth as outcomes of inquiry. For John Dewey (1929, 1938) and C. S. Peirce (Thompson, 1963), truth is a postulated end-state of continued inquiry—a state toward which committed inquirers are moving even though the destination may never actually be reached. For Karl Popper (1968, 1972), truth serves as a regulative ideal—a state we cannot reach with certitude but one we understand well enough to use as a guide to inquiry. Hence, in Popper's view, we continue to falsify our conjectures until we find those that, tentatively at least, resist falsification. These we retain but always with the understanding that they should be reexamined if legitimate objections are thrown up against them.

Philosophers who lean toward a pragmatic theory of knowledge generally regard knowledge as more encompassing than truth. Whereas older epistemologies viewed knowledge as that bit of truth that investigators had discovered and established firmly, the new epistemologies (and postepistemologies) regard knowledge not only as the outcome of investigation but also as that which guides inquiry. C. I. Lewis (1949, 1970), for example, spoke of two phases or moments of knowing—the phase of entertainment and initial interpretation, in which an inquirer uses established concepts to interpret present sensations and pose a problem, and the phase of reinterpreting, verifying, and evaluating the results of inquiry. In the latter phase, an inquirer should have a form of knowledge somewhat more refined and better established than that of the first phase, but this is not always the case. Sometimes even deeper questions arise that cast doubt on the initial concepts themselves. The knowledge that is used to guide inquiry is the product of past inquiry, and the results of present inquiry will guide future inquiry.

The classic debate between Dewey and Bertrand Russell centered on the pragmatist claim that all knowledge and truth are the outcomes of inquiry. When Dewey claimed that even logic must be the outcome of inquiry, Russell scoffed at the notion. "How could one conduct an inquiry *without* logic?" Russell asked. To Russell, the idea of establishing logic by a procedure that in itself required logic was idiotic. But antifoundationalists have tried to show that logic—like all forms of knowledge—develops; that is, more trustworthy forms grow out of inquiries guided by less trustworthy forms. The search for an ultimate beginning, something on which all else can comfortably rest or from which all else will emerge, is a mistake.

An ethical position was at least implied in the older epistemologies. Once a bit of truth was snared and placed in the circle of established knowledge, learners were obliged to accept it. Students could rightly be expected to learn the material so established, and, it could be argued, they had an ethical obligation to do so if they were to be involved in any enterprise for which the material was relevant.

Contemporary cognitivists who hold to the earlier epistemologies generally accept the idea of knowledge as justified true belief. They reject a mechanistic or impressionistic theory of learning, but they do not reject the traditional epistemology. Their emphasis on student activity and understanding is compatible with epistemological emphasis on *justified* true belief; that is, students must be able to give sound reasons for believing X (where X is "true"), if they are to be credited with *knowing* X. Thus one can be a cognitivist without being a constructivist. One can believe, that is, that understanding requires active inquiry and reflection without believing that all knowledge is constructed by individual knowers.

In traditional positions, great emphasis is placed on method because right method plays an important role in justifying belief. When teachers ask, "How did you get it?" they expect students to demonstrate an acceptable procedure. If the procedure is not a familiar one or is one not usually applied to the category of task at hand, teachers press for a justification of method. The widespread reliance on right method in fields such as mathematics induces an ethical obligation in pedagogy. Teachers must be sure that students learn the accepted methods, how to use them appropriately, and—in the best educational situations—why the methods are acceptable.

At the university level, we are all aware of the hegemony of established methods. Most of us find it ethically improper for a professor to allow a student to use a highly controversial method of research unless the student is fully aware of the method's doubtful status. If a well-informed student undertakes such a method, we expect her or him to provide a justification for use of the method. Most of us also feel a strong obligation to be sure that our students understand traditionally accepted methods as well as the rationale for a new one.

The obligation with respect to method raises some interesting questions about precollege mathematics instruction. Piaget (1971) suggested that continuous construction on the part of learners in mathematics would lead steadily toward more mathematically acceptable results. Contemporary constructivists have been careful to elaborate on that claim by describing the kinds of learning environments needed to move students toward ever more powerful constructions. Turning students loose "to construct" will not in itself ensure progress toward genuinely mathematical results. Teachers must ask questions that challenge ill-formed hypotheses and weak conjectures; they must pose new problems that require the revision of old constructions; and, sometimes, they simply must show how things are done. In the last case, wise teachers take note of their own decision to tell or show and watch for later opportunities to encourage construction.

Constructivist teachers believe that students *inevitably* perform construction—some flimsy and indistinguishable from rote learning, some powerful and highly generative—and we cannot be sure what kinds of constructions students make when we demonstrate a technique by showing or produce a solution by telling. Some students will make their own powerful constructions even in such situations; more will perform the weak constructions consistent with associative learning. Because they recognize that weak constructions may follow their telling, constructivist teachers try to assess what their students can do with transmitted knowledge.

Let us consider an example in some depth. We can play this example out in both a classroom scenario and a conversation among teachers. Teacher A describes an approach to teaching students how to complete the square. She tells her class: "Today I'm going to show you a technique. It will be like a game." Then she shows them how to fill in the blanks in the equation $X^2 + 2X + \_\_\_\_ = (\quad)^2$. Teacher A has the class practice several sample problems. Then she explains how the technique is often used to transform expressions and invites her students to rewrite $X^2 + 2X =$ as a perfect square and whatever extra term is needed to maintain the integrity of the original expression. Students try transformations of the sort $X^2 + 2X = (X^2 + 2X + ?) - ?$ So far there has been no real invitation to construct. But now teacher A poses variations: What would you do with $X^2 + 2/3X + ?$ Many students will figure this out. Then teacher A will ask: What do you do with $2X^2 + 4X + ?$

In the teachers' conversations, teacher B may protest. "Wouldn't it be more meaningful to start with an equation such as $X^2 - 2X = 8$ and show how the method can be used to solve an equation? Why the emphasis on gamelike technique?" Teacher A tells a little story. When she was in high school, she encountered the technique of completing the square only in developing the quadratic formula. In college mathematics, she finally met the method again—in expressing the equation of a circle in "standard" form—but hardly recognized the technique. Maybe, she suggests, students will profit by concentrating on a technique for transforming expressions. Teacher A goes on to explore variations in telling and constructing. "Sometimes," she says, "it makes sense to start with a problem and ask students to invent a technique. But sometimes it's useful to demonstrate a technique and then seek meaning for it. Why not?"

Teachers A and B then go on to explore a host of possibilities for student construction. It is clear to both of them that "telling" can be skillfully interwoven with opportunities for students to make powerful constructions. Teacher A says that her students would not allow her to proceed for long without demanding a search for personal meaning. "They quote me to myself! 'This must be going somewhere,' they say, 'we know mathematics isn't just filling in the blanks!'" Both teachers consider the nature of the topic at hand as they debate the wisdom of starting with telling or posing a problematic situation. Both agree that it is disastrous to settle for "telling" from start to finish.

In all of their planning, teachers A and B are keenly aware of their ethical responsibility to encourage students to discuss what they are doing. Particular opportunities to construct will not induce genuine construction from all students. Therefore, teachers A and B watch over their students carefully—treasuring each insight and encouraging every promising effort. A powerful construction at any stage in the treatment of a topic may bring it into the full light of understanding.

Teachers who believe that students must construct their own knowledge are ethically obligated to provide opportunities for them to do so, but such opportunities can occur at all stages of instruction. A teacher might, for example, pose a problem and, with no instruction at all, ask students to suggest ways of approaching it. In another situation, a teacher might give detailed instruction followed by more or less rote practice and then ask: "Now what would you do with this interesting variant?" Somewhere along the line, students have to grapple with the material in genuine acts of construction—that is, strong acts rather than the weak ones characteristic of rote learning (Confrey, 1990)—but they do not have to do everything from scratch.

Constructivist researchers share the ethical obligation to continued inquiry. We must resist the temptation to adopt constructivism as an ideology and, instead, use it to investigate problems posed not only within constructivism but from without as well. Sometimes these problems arise directly out of objections raised to constructivist recommendations, and they should be the objects of open, continued inquiry. Thorough study of what it means to construct, the nature of constructive acts, and the conditions under which construction takes place is an ethical as well as a pedagogical and epistemological imperative.

I will say more about the ethical aspects of pedagogy later, but here I will conclude by noting that children, as well as older students and practitioners, should be aware that the methods they adopt are subject to the scrutiny of a community. They should accept the obligation to state their purposes, explicate their procedures, and justify their results, although justification need not fall along traditional lines. For us teachers, the purposes, procedures, and reasons that we accept will depend not only on our epistemological perspective but also on our ethical position. If we are operating from an ethic of caring, we may, for example, accept a reason that is epistemologically weak from student A and reject the same reason from student B because we feel that the reason is sufficient for student A's purposes but not for student B's. This observation requires further discussion of constructivist pedagogy and ethics.

Constructivist Pedagogy

Constructivism, like other cognitive positions, puts great emphasis on the activity of learners. Indeed, even the National Council of Teachers of Mathematics (NCTM, 1989), without adopting an explicitly constructivist position, recommends the active engagement of students in building an understanding of mathematical concepts and processes of thinking. But what will secure the active engagement of students?

Here we encounter a great gulf between cognitive and dynamic (or motivational) psychology. Often cognitivists depend heavily on the nature of the task to secure engagement. Some writers believe, for example, that real-world problems are more interesting to students than typical school problems. Although this may be true for some students and some problems, it is clearly not true for all students and all problems. I can remember from my own school days how bored I was by problems involving the reading of gas and electric meters and even by those involving check writing, budgets, and interest payments. Perversely, I was totally absorbed in the construction of a quick way to simplify complex fractions.

Even those mathematics educators who are cognitivists or constructivists pay too little attention to the purposes of students. Having given up a stimulus-response theory of learning, they still have not explored deeply enough the internal nature of motivation. Students are motivated from within; they have their own needs, interests, desires, and purposes. Just as constructivist teachers need to know how students think about mathematical problems, so they also need to know why students like or dislike mathematics, what purpose mathematics plays in their lives, and what other goals they are trying to accomplish through present activity in mathematics.

The theme of this book is relating school mathematics to real mathematics, and it is surely an important theme, but there are complications in trying to define what we mean by "real" mathematics. The idea is intuitively clear: Real mathematics is mathematics actually used in the world. Surely this includes mathematics used in both adults' and children's activities. But not everything that interests one adult will interest another, and much of what interests adults will fail to interest children. If we use as a criterion of real mathematics that the topic or problem chosen must be one actually engaged by people in some recognized occupation, and we decide to build our curriculum on such topics and problems, then we certainly construct a different picture of the mathematics curriculum. But I do not see how we necessarily change the motivation of students by doing this. If my purposes, as a student, have nothing in common with those of a biologist, I may be just as bored by a "real" biological problem as I am by a decontextualized exercise in simplifying radicals.

It seems to me that we cannot rely on the task alone to induce mathematical thinking, although some tasks are richer than others in possibilities. There are several current movements in mathematics education that make this mistake or a similar one. Suppose, for example, that we approach division by taking children to a supermarket at which we will try to figure out per-unit prices. This is "real world," but it is also highly artificial—even deadly boring—unless students really have some concern about the purchase of supermarket items. Most do not. They are not the least concerned about saving pennies here and there. This situation in which division naturally arises is not one in which most students place themselves.

In contrast, consider an activity that a cousin and I engaged in when we were both about ten. We played dice baseball, and we were seriously involved in this enterprise for about two years. Our parents bought each of us a small metal file box in which we kept records of our teams and players. He had the National League and I had the American League. (Of course, these teams did not play each other in the regular season, but ours was, after all, a dice game—not the real thing.) We kept the averages of every one of our players. Our method was to divide the number of times at bat into 1,000 and then multiply by the number of hits. In mathematical symbols, we found our averages by converting a fraction such as 9/27 into 333/1,000. Having just completed fourth grade, we knew little of decimals, but we were good at dividing and multiplying. This game made us whizzes! As the first summer wore on and our make-believe/real players accumulated more and more at-bats, we decided to compose a file card of conversions. We divided everything from 2 up to about 50 into 1,000. Then all we had to do was consult the chart of quotients and multiply by the number of hits.

This game was "real life" to us, but it might not strike other children as at all interesting, and plenty of children would resist the task we created happily for ourselves—performing a series of divisions in order to save ourselves work later. It was not just baseball averages that induced our arithmetical activity; it was our own personal interest in baseball and dice baseball in particular.

The very idea of depending on the tasks themselves to induce motivation is antithetical to constructivist beliefs. In a fundamental sense, children construct their own real worlds, and there is no such thing as a universally real problem. For some children (I was one), school *is* the real world — more important and vital than any other world. These children are motivated to do whatever arises in schools, although, of course, they will be more excited by some topics than others.

Nothing said so far suggests that we should give up trying to relate school mathematics to real mathematics. Attempts to revise evaluation methods, such as those reported by de Lange, are valuable if we want to find out whether students can apply mathematics as people do in certain work settings. But a curriculum that produces such proficiency will not necessarily produce greater motivation in students. Similarly, studies such as Carraher's in real-life settings can tell us much about children's thinking in mathematics, but the transfer of such tasks into the school curriculum will destroy their realness for most students. Thus, although there are potential rewards in the present project, increased student motivation is probably not one of them.

Another current notion that makes a similar error is that of the cognitive apprenticeship — the idea that a student will learn from a teacher as an apprentice does from a master, and that if students work with masters who model mathematical thinking and produce fine mathematical products, the students also will learn to do so. What is overlooked in this approach is that apprentices usually choose their line of work. They apprentice themselves willingly to learn a field they want to enter. Sometimes, of course, children simply fall into apprenticeships with a parent or other relative, but in these cases it is often love and admiration for the parent that motivates the apprenticeship. There are no doubt cases on record where unwilling apprentices have succeeded in occupations they would not have chosen, and there must also be cases of highly talented apprentices who overcame the handicap of working with inept and hated masters. But by and large there is a special relation between master and apprentice. It is not just the master's example that counts. The relation between master and apprentice and the apprentice's desire to learn the trade of the master also matter. One does not become a cognitive apprentice in a field in which one has little interest, nor does one choose a master one hates. In the discussion of caring, I will review both of these areas — the caring relation of teacher and student and the need to care for the subject itself.

A supposition sometimes voiced explicitly by advocates of the cognitive apprenticeship is that students should learn from the master how to think like a mathematician. But unless we mean by this simply that students should learn to think clearly and effectively in situations involving mathematical objects or operations, this supposition or recommendation is questionable.

Lots of people other than mathematicians use (and even invent) mathematics effectively. They learn to do so because they have some purpose for doing so. Later, another purpose may guide how they use mathematics. To suppose that either the proper task or the proper model will necessarily induce mathematical thinking is to overlook the fascinating variation in internal motivation. Constructivists, believing that people necessarily construct their own knowledge, should recognize that they also construct for their own purposes, and those purposes affect the style, depth, rigor, and perseverance of the investigation.

Caring and Pedagogy

Most educators recognize the fundamentally moral nature of teaching. Teaching is a moral enterprise—one characterized by moral ends and means. Educators need more than a cognitive perspective; they need an ethical view as well. I will argue that caring and constructivism are eminently compatible and that, without such an ethic, constructivism risks abandonment as just another method or set of methods to be judged in more or less standard terms.

Caring supplies a moral outlook for constructivism. Its emphasis on needs, relation, response, and responsibility suggests attention to all of the needs and cares of students—not only to how they think in mathematics. Both caring and constructivism require receptivity on the part of teachers and appropriate forms of response from students. Caring reminds us that students may need to learn how to respond; many may even need to learn how to be cared for. A consideration of caring reminds us that there are moral as well as mathematical goals in teaching. As educators, we want to produce competent, caring, loving, and lovable people.

I have described caring as relational (Noddings, 1984). A relation or encounter may be called *caring* if both parties—carer and cared-for—contribute appropriately. The consciousness of the carer is characterized by engrossment or nonselective attention. In this ultimate act of receptivity, the carer strives to receive what the other is feeling, undergoing, or trying to accomplish. Simone Weil (1952) says of this form of attention: "The soul empties itself of all its own contents in order to receive into itself the being it is looking at, just as he is, in all his truth" (p. 59). Part of what the teacher receives in an act of caring is a sense of what the student cares about, and because the student matters to the teacher, so does what he or she cares about.

But attention or engrossment must be accompanied by motivational displacement—the desire to meet the other's need, help in her project, contribute to her growth (Mayeroff, 1971). In moments of genuine caring, our motive energy flows toward the purposes and projects of the cared-for. If the purposes of the cared-for are compatible or can be made compatible with our own, we already have an energizing response from the cared-for. When, for example, students want to learn what we want to teach, our teaching can be wonderfully enlivened. But when their purposes are different from ours, we must decide whether to empower them in their purposes or to persuade them to accept ours as their own. This is a moral problem that is too often construed as a mere problem of pedagogical strategy. I will return to it shortly.

If a relation is to be caring, the cared-for, too, must contribute. He or she does so by responding to the efforts of the carer. The response need not be one of gratitude or even direct affirmation. It might be enthusiastic or even reasonably happy pursuit of agreed on projects. When a student's response shows that the efforts to care have been received, the act of caring is completed, and the relation may properly be called caring.

In equal, long-term relations, carer and cared-for regularly exchange positions. We expect relations between spouses, friends, and colleagues to be of this sort. But teaching and parenting are, by their very nature, unequal relations. One party serves more or less continuously as carer. In these relations, the response of the cared-for is crucial, because that response is his or her almost total contribution to the relation. Therefore, helping children learn how to discern and respond to caring, and to elicit further efforts at caring, to articulate needs is a critical task in teaching.

Now let us consider the great moral problem I mentioned earlier—that of deciding what to do when the student's purposes or needs do not match our own. One cannot be the carer in a relation and remain indifferent to what the cared-for is undergoing or striving for, but carers can legitimately respond in a variety of ways. At one extreme, we find people like A. S. Neill (1960), who would not teach mathematics to any child until the child expresses a need or desire for it. At the other, we find people like Mortimer Adler (1982) who feel that caring equally for children requires that we give them all—whether they want it or not—an equal dose of mathematics, English, history, and science.

I think both extremes are hard to justify in an ethic of caring. The permissive stance ignores the possibility that a child—coaxed, inspired, or persuaded by a loving teacher—may actually come to enjoy mathematics or, at least, find it compatible with some of his or her own purposes. The coercive approach depends heavily on the teacher's capacity to convince students that genuine care and concern underlie the coercion. Coercion is always ethically suspect, and we all know cases in which adults have done dreadful things to children "for their own good." A case of caring coercion comes readily to mind in the teaching of Jaime Escalante, the real-life hero of the movie *Stand and Deliver*. We cannot in good conscience recommend that teachers emulate Escalante in the regular use of sarcasm, in putting students down, in throwing them out of class. But what comes through for many of

Escalante's students is his underlying concern and steady care for them as worthy persons. I still worry about those students for whom care does not come through—for those who abandon both mathematics and their own sense of self-worth because they cannot meet their teacher's expectations. I worry also about pernicious side effects in successful students who may equate coercion with care and be badly deceived in the future, or—especially troublesome—may become dependent on a coercive leader, boss, or teacher.

These are deep, difficult ethical problems. Whereas constructivism requires only that we learn how our students are thinking, caring requires us to elicit and listen to how they are feeling, to evaluate their purposes, to help them to engage in self-evaluation, and to help them grow as participants in caring relations. As carers, we need to consider the potential legitimacy of many motives.

Consider one motive that I heard over and over in my years as a high school mathematics teacher. When I asked, "Why are you taking math?" many students answered, "Because it's required for college." They often added, "I wish it wasn't. I hate it." Should I, as a mathematics teacher, have had the same expectations for such students as I had for those who loved math? This is a question that must be answered within a moral framework. Many educators today answer it too glibly, using slogans that have a nice ring to them: All children can learn! Have high expectations for all students!

Such slogans capture something of moral significance, but they need to be analyzed and explicated. Certainly we should not establish our expectations on the basis of race, ethnicity, or gender. Just as certainly, we should help all students to identify their strengths and build on them. We should help all of them to evaluate their own work and to appreciate quality in their own work and that of others (Glasser, 1990). But this does not mean that we should have uniformly high expectations for all individual students in mathematics.

When we care for students, we listen to what they say and to what interests them. It may not be mathematics that interests them. It is not the obligation of mathematics teachers to make everyone love mathematics. Secondary school is a highly coercive institution, one that forces students to expend considerable energy on matters they find dull and irrelevant to their purposes. In opposition to those traditional moral perspectives that place a moral obligation on students to accept the values authoritatively established, an ethic of care requires teachers and students to construct values cooperatively. A mathematics teacher has responsibilities to several entities: first, to the student; after that, to the school, to teaching as a profession, and to the mathematical community. With students whose primary interests are not in mathematics, teachers properly negotiate goals and levels of proficiency that will allow them to certify to the relevant institutions that a satisfactory job has been done. Establishing exactly what is meant by "satisfactory" will always call for professional judgment.

Further, just as the researchers represented in this book are trying to relate school mathematics to real mathematics, we should try to relate the mathematics we teach in schools to the other subjects students study. Neither attempt at forging connections will solve all motivational problems, but both will increase the possibility that students will find reasons for studying mathematics.

One way to connect with other subjects is through biography and existential questions—questions that are central to human life. When students are introduced to rectangular coordinates, they should hear about Rene Descartes. Many mathematics texts include pictures of great mathematicians accompanied by a few lines of biographical information, but this brief exposure—if the teacher even provides it—is hardly enough to intrigue students or to bring Descartes and his questions to life. Students should hear that Descartes led a colorful life, that he was a fashionable dresser and engaged in swashbuckling adventure. He was adept at swordplay and ready to draw his sword when rowdies insulted him or the lady he was escorting. Dressed like one of the three musketeers, Descartes cut quite a figure (Bell, 1937).

Besides biographical information that may interest many students, Descartes' philosophical work should be discussed. Teachers rarely find opportunities to present and discuss great existential questions in today's schools, but mathematics classes can provide natural settings for such discussion. One question that deeply concerned Descartes was whether it can be shown that God exists. Surely, students—both believers and unbelievers—have wondered about the existence of God or gods. Descartes revived and polished St. Anselm's ontological proof, which is based on the idea that God is conceived as the perfect entity, and for an entity to be perfect it must exist. The idea of perfection carries with it, of necessity, the provision that whatever is perfect must exist. Otherwise, it would not be perfect. Can students find anything wrong with this line of reasoning?

They should hear—after they have given it considerable thought—that another great philosopher, one who made significant use of Euclidean geometry in his thinking, found a fatal flaw in the ontological argument. All it seems to prove is that a *concept* of perfect entity or perfection exists—not that the entity to which perfection is ascribed exists.

Mathematics teachers could also prompt their students to ask their biology teacher to discuss the teleological argument for the existence of God and their physics teacher to tell them about the cosmological argument. (Perhaps they should warn their colleagues first!)

When we care for our students and the things that interest them, we should be led to share with them the lives and interests of mathematicians and well-known users of mathematics. Mathematicians, like the students themselves, have had interests other than mathematics. When probability is studied, students should hear about Pascal—not only about his interest in games of chance but also about his consuming interest in existential questions, especially the same one that interested Descartes: Does God exist? Pascal did not try to prove God's existence. Instead, he offered a well-known wager: If you bet on God's existence, and God exists, what do you stand to gain? If God does not exist, what do you lose?

Mathematics is filled with opportunities of this sort. When students are introduced to truth tables, they should hear about Wittgenstein and how he retreated from scholarly life to manual labor for years because he thought the most important things in life could not be proved or even stated but had to be shown in one's own life. Students may find validation for their own concerns in stories such as these (for many more examples, see Kung, 1980). In all this work, we are relating school mathematics not just to real mathematics but to real life itself, in all its depth and wonder.

Even "real" mathematics, as it is described by contemporary mathematicians, goes well beyond direct applications in adult life and connects with other school subjects. Marjorie Senechal (1990) describes a myriad of uses for art in mathematics. Children can construct various solids, picture various shapes, learn projective techniques, perform transformations on computerized objects, and study the natural world of shapes. Senechal uses terms that some students never encounter in math classes: dissection, shadows and lenses, drawing, image reconstruction. She discusses soap bubbles, kaleidoscopes, crystals, snowflakes, fractals, molecules, and tilings. On many of these topics, math and art teachers could collaborate to the advantage of both and of their students as well. Students whose major interest is in something other than math might find a surprising connection to their own central interest.

It has been my experience that students understand and appreciate the kind of caring just described. Teachers using an ethic of care enter into genuine dialogues with their students. They talk about and encourage legitimate student interests, and they talk about their own interest in the subject matter they teach. They also convey to students a sense of multiple responsibilities and how these must be balanced continually. Students learn much more than mathematics from such teachers as they discuss the demands and interests of individual existence and of various institutions and communities.

An ethic of care is directed at creating, maintaining, and enhancing caring relations. From this perspective, it is essential that teachers and students have time to establish relations of care and trust. Such relations are best created over a period of several years; many European schools already use a system that keeps teachers and students together for several years. But we must first legitimize relational goals. In our present system of schooling, the greatest emphasis is on academic goals. Even these, as constructivists have complained, are far too narrowly construed. However, it is not enough simply to describe academic goals more broadly. Teachers and students need time to develop the kind of relations that make it possible for teachers to advise, persuade, and negotiate intelligently. Teaching is not just a matter of putting information into students' minds, or of eliciting ideas from within, or even of promoting genuine mathematical thinking. It also involves fostering general growth—intellectual, social, and moral—and it involves living together in the kind of relation that prepares students for life-long cooperation and appreciation.

There are, of course, potential academic advantages in establishing relations of care. As the poet Goethe noted, we learn from those we love. Because many of Jaime Escalante's students have detected the love and care that underlie his sometimes heavyhanded strategies, they respond to his exhortations with a high level of effort. Some students may be driven by a newly acquired intrinsic interest in mathematics; some may be responding to a fresh appreciation for their own capacities (Escalante tells them repeatedly, "You are the best!"); some may be working toward occupations that require mathematics and for which they now have realistic aspirations; and some may be responding simply (magnificently) out of love for their teacher. Teachers who are guided by an ethic of care recognize all of these motives and stay with their students through times of vigorous participation and intervals of low energy and flagging interest.

Just as teachers have a special obligation to modify goals and expectations for those students whose major interests lie outside mathematics, so they have special, but different, obligations with respect to mathematically talented students. In this area, our earlier discussion about epistemology and ethical commitment becomes especially relevant. Students who want to enter some subset of the mathematical community must be introduced to the mores of that community. They can be held to higher and more stringent criteria of mathematical justification than other students, and they can enter into meaningful apprenticeships. To care for these students entails helping them to care for mathematics—not only to like it but to understand its rigor and beauty, to know something of its history and current problems, to emulate its patterns of thinking, and to accept its standards of truth. Too often teachers sacrifice the pursuit of these lovely goals because they mistakenly suppose that they must have the same uniform (high) expectations for all students. As a result, they serve no students well.

Constructivism and caring, as I said at the outset, are peculiarly compatible. Constructivism holds that knowers necessarily perform constructions in building their own knowledge. Caring reminds us that people have their own motives for performing constructions and have many goals that have higher priority than academic ones. Caring constructivists are prepared to work sympathetically with a wide variety of student motives, to stay with their students through positive and negative experiences, and to seek consistently to promote the general growth of their students.

I will close this discussion of constructivism and caring with a few words on evaluation. Should the "extra" material on biography, art, religion, and the like be included in formal evaluation? As our students always ask, "Will this be on the test?" Of course this material should appear in whatever mode of evaluation is used. It is important, is it not? But it should appear, as should challenging mathematical problems, as an opportunity. After demonstrating their competence at whatever mathematical manipulations and understandings are considered essential, students should be allowed to choose from a wide range of questions which ones they will answer to show their own special expertise. In this way, we encourage the exercise of a full range of human intelligences (Gardner, 1982, 1983).

Conclusion

Constructivists believe that knowers must construct their own knowledge, but the knowledge produced by a given set of constructions may not be as adequate, accurate, or powerful as that produced by further construction. For that reason, constructivism requires, as the early pragmatists recognized, a commitment to continued inquiry. This commitment is essentially ethical, because it requires us to decide what the purpose of our investigation is and to tailor it to fit our purpose. In working this way, we must be aware of the limits on our claims and convey them honestly to colleagues.

Constructivist pedagogy requires a similar ethical commitment. Understanding and accepting student purposes, we ask different questions of different students and urge them to design their investigations so that they are adequate for their own well-considered purposes.

Constructivist pedagogy requires the receptivity and responsiveness characteristic of an ethic of care. When we care for our students, we take an interest in their purposes as well as our own. Care thus leads us to connect school mathematics not only to "real" mathematics but also to real life in all its breadth and wonder. Teachers who would care in this way obviously need time to develop relations of trust with their students, and they need time to converse—to share and to listen. In an atmosphere guided by constructivism and care, much more is learned by both students and teachers than can be prescribed at the outset. Most important is that both become competent, caring, loving, and lovable people.

References

Adler, M. J. (1982). *The Paideia proposal*. New York: Macmillan.
Bell, E. T. (1937). *Men of mathematics*. New York: Simon and Schuster.

Confrey, J. (1990). What constructivism implies for teaching. In R. Davis, C. Maher, & N. Noddings (Eds.), *Constructivist views on the teaching and learning of mathematics* (pp. 107-122). Reston, VA: National Council of Teachers of Mathematics.

Dewey, J. (1960/1929). *The quest for certainty.* New York: G. P. Putnam's Sons (Capricorn Books edition).

Dewey, J. (1938). *Logic: The theory of inquiry.* New York: Henry Holt.

Gardner, H. (1982). *Art, mind and brain.* New York: Basic Books.

Gardner, H. (1983). *Frames of mind.* New York: Basic Books.

Glasser, W. (1990). *The quality school.* New York: Harper & Row.

Kung, H. (1980). *Does God exist?* Garden City, NY: Doubleday.

Lewis, C. I. (1949). Experience and meaning. In H. Feigl & W. Sellars (Eds.), *Readings in philosophical analysis* (pp. 128-145). New York: Appleton-Century-Crofts.

Lewis, C. I. (1970). The pragmatic element in knowledge. In J. D. Goheen & J. L. Mothershead, Jr. (Eds.), *Collected papers of Clarence Irving Lewis* (pp. 240-257). Stanford, CA: Stanford University Press; also, Review of John Dewey's The quest for certainty.'" In *Collected papers* (pp. 66-77).

Mayeroff, M. (1971). *On caring.* New York: Harper & Row.

National Council of Teachers of Mathematics. (1989). *Curriculum and evaluation standards for school mathematics.* Reston, VA: Author.

Neill, A. S. (1960). *Summerhill.* New York: Hart.

Noddings, N. (1984). *Caring: A feminine approach to ethics and moral education.* Berkeley: University of California Press.

Piaget, J. (1971). *Biology and knowledge.* Chicago: University of Chicago Press.

Popper, K. (1968). *Conjectures and refutations.* New York: Harper & Row.

Popper, K. (1972). *Objective knowledge.* Oxford: Oxford University Press.

Senechal, M. (1990). Shape. In L. A. Steen (Ed.), *On the shoulders of giants* (pp. 139-181). Washington, DC: National Academy Press.

Thompson, M. (1963). *The pragmatic philosophy of C. S. Peirce.* Chicago: University of Chicago Press.

Weil, S. (1952). *Waiting on God.* London: Routledge & Kegan Paul.

CHAPTER FOUR

The Reality of Negative Numbers

ROBERT B. DAVIS   CAROLYN A. MAHER

Abstract

Business people, mathematicians, mathematics educators, and psychologists have all come to pay attention to the mysteries of negative numbers. In Chapter 5, Terezinha Nunes looks at the way the ideas of negative numbers are developed within school and outside of school. Because there are many different views of subtraction, additive inverses, and negative numbers, it may be worth a brief consideration of some of these differences.

Few, if any, topics illustrate the interplay of "reality" in its various manifestations with the formal structure and imagery of mathematics better than the ideas associated with negative numbers. In this single topic we see the contrasting possibilities of meaningful versus meaningless symbol manipulation, the role of "situated" mathematics where a real context may shape our thinking, and the developmental nature of conceptual growth, both within the society at large and within the individual student.

"Two-Attributes"

In order to understand how humans think about negative numbers, it is important to distinguish between what might be called the "two-attributes" conceptualization of "gains" and "losses" and the quite different "single-attribute" concept. As we shall see, the single-attribute notion is more sophisticated and more powerful, and probably appears later, both in the life of the individual and in the history of the society.

Let us look first at the question of attributes in a simpler instance. Temperature, being more concrete, can be helpful. In elementary school science, it is well known that many children, asked to put their hands in tubs of water at different temperatures, show us a notion of attributes that is different from that of adult scientists. Suppose, say, that the water in one tub is warm and the water in the other is ice cold, with unmelted ice cubes floating around in it. If we ask children "which tub is warmer," many children will refuse to answer what they regard as a silly question. The water in one tub is not warm at all! Quite the contrary; it is ice cold!

The children who respond in this way are, of course, showing us that for them "being warm" is one kind of attribute, and "being cold" is something entirely different. It is as if we showed them a magenta crayon, then let them hear a loud noise, and asked them "Which is redder?" Surely the magenta crayon is at least a little bit reddish, but the loud noise is not red at all, so one is unable to rate it on a scale of "reddishness." There can be no comparison.

But of course this is not how scientists or mathematicians view temperature. Everything has *some* temperature, it is just that the temperature of some things is higher than the temperature of others. Given *any* two objects, we can meaningfully ask, "Which one has a higher temperature?" or, in other words, "Which one is warmer?"

If you regard "warm" as one kind of attribute, and "cold" as a different attribute, you are taking what we call the "two-attribute" approach. Historically, this is the earlier, less sophisticated conceptualization. It also comes first in the life of an individual child. Only later does one create the more powerful "single-attribute" version, which sees everything laid out along a single scale that we might call *temperature*, instead of using two separate scales, one for *coldness* and the other for *warmth*.

Exactly the same developmental sequence takes place also in the case of negative numbers. Here, too, ontogeny recapitulates phylogeny. What was true for the society is true also for the individual child; the sequential order of conceptual development through which humankind passed is the same sequential order of conceptual development through which each individual student will pass—except that, unfortunately, not every student will reach the more mature conceptualizations. Many adults never do.

Students (or adults) reveal that they are still mired in the two-attribute conceptualization when they say, for example, "Oh, I see! That's a loss of negative four." These students are thinking in terms of two separate scales, one for "gains" and the other for "losses." The imposition of *negativity* on one of these scales is essentially redundant. (Of course, it is also wrong.) In this form, the true power of positive and negative numbers cannot come into play.

If, instead, we wish to express the ideas of gains and losses in one-attribute form, we should say either "a gain of negative four" or else "a loss of four"—or even "a loss of positive four." Only one attribute is involved—the idea of change—and the sign of the change alone distinguishes a gain from a loss. "Cancellation" becomes automatic, and one can always compute the new balance by "adding" together the old balance and the change:

$$A_{new} = A_{old} + \text{change.in.}A$$

Why don't we need to distinguish "addition" from "subtraction"? After all, in some cases the change may be a gain, and in other cases it may be a loss. Aren't we making some kind of inappropriate assumption when we treat every case by *adding* the change?

No, we are not, and this is the power of the one-attribute conceptualization. We can (and, in fact, *must*) always *add* the change, leaving it for the *sign* of the change to determine whether our balance is growing or shrinking. This is, of course, the essence of algebraic addition, which is quite different from the addition of usual elementary school mathematics (for a more detailed discussion, see Davis, 1984, pp. 165–170).

Historically, the Chinese had the two-attribute version of negative numbers from ancient times (far earlier than Europeans), using red to indicate positive numbers and black to indicate negative numbers (Smith, 1925; Boyer, 1991; Crowley & Dunn, 1985). In this form, of course, cancellation was not automatic, as it is in the one-attribute version.

When did the "one-attribute" concept become fully developed? Historically, much later, especially in Europe; the real generality of the idea may not have occurred before Hudde, in 1659 (Smith, 1925). Even mathematicians of the sophistication of René Descartes did not accept negative numbers as entirely legitimate; indeed, Descartes spoke of "true" solutions of equations and "false" solutions, the former in the case of nonnegative roots, the latter in the case where the alleged root was negative (Boyer, 1991, p. 345). Summarizing the situation, Crowley and Dunn (1985) write:

> As late as the sixteenth and early seventeenth centuries, many European mathematicians would still not accept . . . [negative numbers], or, if they did, would not accept them as roots of equations. (p. 252)

Agreeing with this assessment, Jones (1954) makes the further point that "Popular acceptance waited still longer for simpler and more common applications and for graphical representations" (p. 41). "By the eighteenth century," Crowley and Dunn (1985) write, "negative numbers were universally accepted" (p. 253). Yet it was only as recently as the nineteenth century that the infamous "negative times negative is positive" rule was fully understood (Crowley and Dunn, 1985). Even Euler, one of the true giants in the history of mathematics, had had no good explanation for this rule, although he accepted it as apparently correct.

In reading Nunes's discussion of negative numbers, it is important to keep in mind the distinction between the two-attribute version and the more powerful one-attribute concept. Only the latter can truly be called "the idea of negative numbers," although the more primitive two-attribute version is undoubtedly a first step in the right direction.

Symbols: Meaningful or Meaningless?

Few questions have bedeviled the teaching of mathematics more than the distinction between manipulating meaningless symbols according to mysterious rules (or should one say "manipulating mysterious symbols according to meaningless rules"?), versus dealing with symbols that have meaning and basing one's thinking not on the symbol but, rather, on the meaning—in effect, regarding each symbol as a window through which one looks at some particular reality. Probably everyone who becomes highly skilled at mathematics must ultimately be able to deal with symbols in each of these ways, choosing whichever method is appropriate for the task at hand.

We see both alternatives at work in the historical development of negative numbers. The "meaningless" use of symbols played an important role because many rules became known in the case of positive numbers, such as

$$(a + b)(a - b) = a^2 - b^2$$

and in certain instances these rules gave an indication of how to assign a value to products of negative numbers. Hence one had, for a time, the spectacle of mathematicians who did not believe that negative numbers had any kind of existence or legitimacy but who could, nonetheless, carry out computations with them. This must have been, in one sense, the apotheosis of meaningless manipulation of symbols. It must also have been exciting and therefore should not be confused with meaningless rote rules that nowadays are sometimes told in class to unwilling students.

"Situated" Mathematics

Probably most readers feel, as we do, that, at least where students are concerned, symbols should usually be introduced in a meaningful way. In recent times it has become fashionable to speak of this as *situated* concept development, although this idea of *situatedness* could hardly be said to be new (cf. Davis, 1989; Brown, Collins, & Duguid, 1989).

Apparently the Greeks also held to this high valuation of specificity. For them the *meaning* of a symbol was paramount. Consequently, in Greek mathematics there is no suggestion of allowing negative numbers to stand by themselves; instead, one has only (as Jones, 1954, expresses it) "subtractive quantities which [must] always be associated with larger additive quantities." (Note also the Greeks' adherence to the "two-attribute" notion of negative and positive.)

The place where meaning was most supportive of the basic ideas was, of course, in *business*. Fibonacci, writing in 1202, followed an earlier Arab custom of "paying no attention to negative numbers" (Smith, 1925, p. 258). A few years later, however (specifically, in 1225), he interpreted a negative number as a *loss* instead of a gain (Smith, 1925; Fraenkel, 1955). This may still have been within the framework of a two-attribute conceptualization but was nonetheless an important step forward, in that it was beginning to seem possible to give sensible meanings to the previously "silly" symbols for negative numbers (for a further discussion of real-world meanings for negative numbers, see, e.g., Kline, 1959, pp. 50–52).

Notation

Let us now jump from historical notions of earlier centuries into the kind of abstract axiom-based system that twentieth-century mathematicians believe algebra to be. There are different versions of today's algebra, but the differences are not fundamental.

This modern version of algebra has several *binary operations*, one of which is called *subtraction*. By saying it is a "binary operation", we mean that it accepts as inputs an ordered pair of numbers; thus, "3 − " would not make sense within this meaning, nor would " − 3"; a description of the inputs requires that there be two of them, as in "5 − 3", and that we pay attention to their order, so that "5 − 3" does not mean the same thing as "3 − 5."

There are also *unary operations* (which, as one might guess, means that the operation accepts only one number as an input). One very familiar unary operation, available on most calculators, is the "square" function. To use it, one gives the calculator only one number as an input. It is thus quite different from the " + " and " − " keys, either of which would require two numbers.

The unary operator with which we are presently concerned is, of course, the operation of *finding the additive inverse*. The additive inverse of positive three is negative three; the additive inverse of negative seven is positive seven.

(The additive inverse of zero is zero.) By definition, for any number x, the additive inverse of x is that number which, when added to x, would produce the answer zero.

There is a third idea that is relevant here, the idea of a *negative* number. A few decades ago, all three of these ideas — the binary operator, the unary operator, and a part of the name of any negative number — were written the same way and even frequently read the same way, so that one might write

$$7 - 3 = 4$$

and read it as "seven minus three equals four"; or, meaning to set down a name for the number *negative five*, one might write

$$- 5$$

and read it "minus five."

Trouble often arose when several of these different meanings needed to be expressed in the same mathematical equation, as in:

$$-(-5) = 5$$

which might have been read as "minus minus five equals five." A quintessential problematic situation occurred when one tried to define the *absolute value* of a number, by writing

$$|x| = \begin{cases} x, \text{ if } x \geq 0 \\ -x, \text{ if } x < 0 \end{cases}$$

Students would read this as "the absolute value of x is x, if x is not minus; the absolute value of x is minus x, if x is minus." They would then frequently object, "But I thought the absolute value could never be minus!" (for a further discussion, see Stallings-Roberts, 1991; Sink, 1979).

In the 1950s, several people (among them Paul Rosenbloom, then at the University of Minnesota, and David Page, then at the University of Illinois in Urbana/Champaign) set out to replace this very confusing notation with something clearer. If you have three different ideas, it might be better to have three different notations. One solution was to write the name of a negative number with a small, raised sign

$$^-5$$

and to read it as "negative five"; to write the binary operation of subtraction with the usual sign, as

$$7 - 3 = 4$$

which would be read as "seven minus three equals four"; and to use a small raised "°" to indicate the unary operation of taking the additive inverse (equivalent to using the "change sign" key on some calculators):

$$°(7) = {}^-7$$

which would be read as "the opposite of seven is negative seven" (or else as "the additive inverse of seven is negative seven"). A more telling example would be

$$°({}^-5) = {}^+5$$

which would be read as "the additive inverse of negative five is positive five." The meaning of this statement would, of course, be that if you start out at negative five (think of it as a point on a number line, if you wish) and add positive five to it, you will end up at zero. "Positive five is the number that you must add to negative five in order to end up with an answer of *zero*." That is to say, "positive five is the *additive inverse* of negative five."

These notations were among the most valuable contributions of the unfairly maligned "new mathematics" of the 1950s, but were not always recognized for what they actually had to offer. For a pedagogy based on imitation, as far too much pedagogy was in those days, notation really didn't make all that much difference. "Understanding" was not a serious goal, and the ubiquitous "minus" notation was probably as easy to imitate as any other.

Of course, it was *not* easy to understand. But the more common goal was merely imitation. If the student *wrote*

$$-(-5) = 5$$

that was what was desired, and nobody asked the student to explain exactly what it all meant. With this notation (and using □ to mean a variable), the definition of the absolute value would now appear as shown in Figure 4-1 (for further discussion, see Crowley & Dunn, 1985; Davis, 1990, pp. 95–96; Davis, 1967, Chapters 4–6).

A Different Point of View

What we have sketched out might be called a mathematician's view of mathematics, and a mathematics educator's view of how to help students to understand mathematics. But mathematicians and mathematics educators are by no means the only people to concern themselves with school mathematics. Piaget, who was neither a mathematician nor a mathematics

FIGURE 4-1 • *The definition of absolute value, written in a notation that distinguishes the different meanings of "minus"*

$$\left|\,\square\,\right| = \begin{cases} \square, & \text{if } \square \geq 0 \text{ is true} \\ {}^{\circ}\square, & \text{if } \square < 0 \text{ is true} \end{cases}$$

educator, dealt with some aspects of the development of mathematical ideas in children. Drawing on quite different traditions, Terezinha Nunes, in Chapter 5, looks at ways in which psychologists have attempted to analyze children's thinking (and, in some cases, also the thinking of adults).

There is no simple way to map the Nunes categories into those of mathematicians. When children think in terms of "having three cookies and eating one of them," they are clearly dealing with one *concrete* instance corresponding to the *abstract* idea of subtraction: $3 - 1 = 2$. This is the mathematician's familiar binary operator, written as "$-$".

But how about the child who is thinking of Mary having eight dolls and Beverly having five. How many more dolls does Mary have? This is a different kind of concrete instance, but it is often taken to correspond to the same *abstract* idea, the binary operator *subtraction*. Some of us might prefer to consider the "How many more . . . ?" questions as *conceptually* mapping into *addition*: $5 + x = 8$. Indeed, it has too often escaped notice that it can also be *solved* as an addition problem, making use, perhaps, of the important idea of *monotonicity*. Although children do not, of course, know this *word*, they very often make use of this *idea*. Indeed, they do so any time they reach into a pocket for change and find that they have come out with less than they need. They do not wonder whether to put some change back into their pocket and then count what is left in their hand, to see if they now have enough. They know immediately that if they have too little, they need to hunt around in their pocket for some more; they *do not* need to put some money back into their pocket. Whenever you can tell whether a candidate "answer" is too small or too large, you have available to you a very powerful method of solution. For the problem $5 + x = 8$, you can try "1". You see immediately that $5 + 1$ is too little, so 1 is too little. Try "5"; $5 + 5 = 10$, which is too large; hence 5 itself is too large. Thus, the right answer must be *more than 1* and *less than 5*. This can lead quickly to an answer — even faster if you use parity and recognize that the correct answer *must be an odd number* because you will need an odd number to add to 5 to get the even sum 8.

Relatively little research has looked carefully at how children actually solve many of these problems. The paucity of work in this area probably results from the common belief that children must be told how to do mathematics; our own observational studies bring this very much into question. Children are very inventive in this area if only they are given a chance to be so.

A second influence limiting research in this area may have been the assumption that what our schools have right now is the only possible curriculum and that the solution methods taught are the only ones possible. In fact, this is not at all the case, and becomes less so every day, as speed and efficiency become less important goals for humans, and understanding becomes more important. Nunes's category of "part of the name of a negative number" is clearly recognizable as David Page's "negative" sign, written small and high.

One might, however, argue with Nunes's conclusion that 3 − 5 "is devoid of empirical reference because there is no possible physical representation for the action of taking five objects from a set of three." Even the world of children is not limited to discrete mathematics; children compare heights of themselves and their friends, which is a *continuous* variable; children are not limited to only the discrete counting of piles of pebbles and the like. The world of many children also includes thermometers and elevators, where any numbering is necessarily based on more or less arbitrary reference points (see also Kline, 1959, pp. 50–52).

Mathematicians would probably not recognize Nunes's notion of "the minus sign as a representation of inversion." Her example of 6 − (+4) is nothing other than the usual binary operation of subtraction; so, too, is her example 6 − (−4). (We must admit to a personal concern that, as far as the analysis of human mathematical thought is concerned, Nunes's novel interpretation may not be a step in a forward direction.) That a child will often *think* about a sequence of actions being taken in reverse, though clearly true, is not evidence that the *operation of concatenating* has itself been altered. What has been changed, in most cases, are the *individual steps* and the *sequential order* of concatenating.

Pedagogical Devices

Of course, one major concern of all of us is this: How can we help students build up, in their own minds, powerful representations of negative numbers? Of the methods that we have attempted (over the years, a great many methods) the most effective have been the "pebbles-in-the-bag" model for the subtraction of nonnegative numbers (as in 7 − 3, or 8 − 12), and the "postman" or "mail carrier" stories for binary operations on signed numbers. The pebbles-in-the-bag model is described in Chapter 4 of Davis (1967), and in Davis (1990), pages 95–96. "Postman stories" are described in Chapters 5 and 6 of Davis (1967). Teachers must use these methods with some care, but when they are used well they can be extremely effective.

Moral

Negative numbers have intrigued and confused some of the greatest mathematicians who have ever lived. Looking at the conceptual underpinnings of this critical portion of mathematics is well worth doing — but do not expect it to be simple. And do not expect that everyone will agree with you.

References

Boyer, C. B. (1991). *A history of mathematics.* New York: Wiley.
Brown, J. S., Collins, A., & Duguid, P. (1989). Situated cognition and the culture of learning. *Educational Researcher,* 18(1), 32–42.
Crowley, M. L., & Dunn, K. A. (1985). On multiplying negative numbers. *Mathematics Teacher,* 78(4), 252–256.
Davis, R. B. (1967). *Explorations in mathematics — A text for teachers.* Palo Alto, CA: Addison-Wesley.
Davis, R. B. (1984). *Learning mathematics: The cognitive science approach to mathematics education.* London: Routledge.
Davis, R. B. (1989). The culture of mathematics and the culture of schools. *Journal of Mathematical Behavior,* 8(2), 143–160.
Davis, R. B. (1990). Discovery learning and constructivism. In R. B. Davis, C. A. Maher, & N. Noddings, (Eds.), *Constructivist views on the teaching and learning of mathematics.* Monograph 4, *Journal for Research in Mathematics Education.* Reston, VA: National Council of Teachers of Mathematics.
Fraenkel, A. (1955). *Integers and the theory of numbers.* New York: Scripta Mathematica.
Jones, P. S. (1954). *Understanding numbers: Their history and use.* Ann Arbor, MI: Ulrich's Bookstore.
Kline, M. (1959). *Mathematics and the physical world.* New York: Crowell.
Sink, S. C. (1979). Understanding absolute value. *Mathematics Teacher,* 72(3), 191–195.
Smith, D. E. (1925; reprinted 1958). *History of mathematics* (Vol. 2). New York: Dover.
Stallings-Roberts, V. (1991). An ABSOLUTE-ly VALUE-able manipulative. *Mathematics Teacher,* 84(4), 303–307.

CHAPTER FIVE

Learning Mathematics
Perspectives from Everyday Life

TEREZINHA NUNES

Abstract

A theoretical analysis of different meanings for the minus sign and an exploratory study of some implications of this analysis are presented. It is suggested that the negative sign has three meanings that cannot be reduced to each other: (1) a magnitude that is marked as negative, (2) the operation of subtraction, and (3) inversion. Negative numbers interpreted as magnitudes implicitly marked as negative (e.g., debts) are easy to understand and operate with. However, the signed, written representation of negative numbers appears to be a source of confusion for students when magnitudes and operations are coordinated; in these situations, the minus sign is interpreted as an indication of subtraction and the marking of magnitudes is interfered with. Data from two experiments on negative numbers are presented to support this analysis, and the educational implications of these experiments are explored.

Mathematics as a formal and specialized representation of concepts often involves situations that are similar to those encountered in learning a foreign language. For example, when we learn the French word for *house* in school,

---

The preparation of this paper was supported by grants from the Brazilian Ministry of Education/CAPES/SPEC/PADCT, which provided me with support during the initial phase of this study while I spent one year abroad, and from CNPq, which provided the support needed for carrying out the studies. Professor Hart of King's College discussed the ideas presented here with me. I am thankful to her for her valuable feedback. Shirley, Suely, Simone, Jorge, Monica, Ana Lucia, and Rossana worked in the data collection and analysis. I am thankful to all of them for their help and to the institutions that made this work possible.

we do not have to learn the concept of *house*. We already know what a house is; all we have to learn is a new verbal representation, the French word *maison*. Similarly, when teaching mathematics, teachers may face situations in which pupils already understand the concept but have to learn a new way of representing it. In this type of situation, the teacher will certainly want to draw on the pupils' previous knowledge when introducing the new representation.

In order to take into account pupils' previous knowledge when teaching mathematics, it is necessary to analyze the situations that give the mathematical concept meaning in everyday life and to find out how pupils understand these situations. This chapter analyzes three different meanings for negative numbers that may be understood in everyday life. It is not intended to provide a mathematical analysis of integers but, rather, to look at situations that make negative numbers meaningful in different ways. Although these situations are different, their mathematical representation and properties are the same. For this reason, in order to understand negative numbers effectively, pupils must relate the representations and properties to all three types of situations and coordinate the three meanings into a coherent system.

The chapter is organized in three main sections. The first section discusses the situations that give meaning to negative numbers. In the second section, two studies are described to support the analysis developed in the first section. A final section reviews the main points presented in the chapter and discusses their educational implications.

Different Meanings for the Minus Sign

The analysis of the different referential situations that give meaning to the minus sign can be approached both in terms of the child's understanding and historically as well. Looking at historical information, Janvier (1985) suggested that numbers were for centuries associated with magnitudes, thereby relying on an implicit model of number as referring to a potential collection of units. Such counting numbers are a subset of the natural numbers. According to Janvier, negative integers were devised to solve equations of the type $x + a = b$, where a and b are integers, and $a > b$,[1] long before the number line with negative positions emerged as a standard interpretation of negative integers. The solution to this equation involves finding the result of $b - a$, which is devoid of empirical reference because there is no possible physical representation for the action of taking a objects from a set b when $a > b$. Janvier further notes that negative numbers "were entities obeying rules with respect to arithmetic operations and vaguely (at times) associated with debt" (Janvier, 1985, p. 135).

This historical account should not lead to the conclusion that children must learn negative numbers as formalizations, studying their properties before they can anchor them in situations that make them meaningful. In the next paragraphs three types of meaning for the minus sign will be discussed. These meanings provide categories for the classification of everyday situations involving negative numbers.[2]

1. *The minus sign and the operation of subtraction.* The first meaning that children can bring to the minus sign when it is taught in school is related to the idea of "taking away" (see, e.g., Brown, 1981; Carraher & Bryant, 1987). However, in phrases like "five bricks take away two," which are represented in mathematics as "5 − 2," *both 5 and 2 are natural numbers (or counting numbers) and refer to magnitudes.* The minus sign in 5 − 2 has to do with the action of taking away, an action that is seen by the children as independent of the number of bricks to which it is applied. *Two* refers simply to a magnitude, the amount of bricks; it remains in the field of natural numbers. The minus sign refers to the operation performed. The question of a negative number does not emerge here. This independence between the number and the operation performed was clearly illustrated in a study by Hughes (1986). Children were asked to represent on paper "5 bricks take away 2." Most often, children simply represented the numbers. When "take away" was represented, however, it was indicated through drawings of hands that were to move the bricks about.[3]

Take away is only one of the situations in which the minus sign stands for the operation of subtraction. The meaning of subtraction as an operation goes beyond this simple situation and develops during the first four years of school. This development will not be discussed here because it has been widely analyzed in the literature (see, e.g., Riley, Greeno & Heller, 1983; Vergnaud, 1982; Carpenter & Moser, 1982).

2. *The minus sign as a mark intimately connected with the magnitude that follows it.* A second meaning for the minus sign was indicated by Janvier (1985), who referred to a "vague association" between negative numbers and debt. Measuring debts seems to constitute a situation that gives negative numbers a different sort of meaning from the one just described. In contrast to the foregoing situation, when the minus sign is used to indicate debt, *it belongs with the number* and does not describe an operation carried out on a natural number. The number itself is considered negative even if the minus sign is not put on paper when people write down their debts.[4]

How can it be demonstrated that debts are quantities marked as negative in the absence of formal representation through the minus sign? From a psychological viewpoint, this can be demonstrated by looking at the inferences made by subjects when debts (a negatively marked magnitude) are added to income (a positively marked magnitude measured by the same standard, money). If a negatively marked magnitude is added to a second one that is positive, the result is not the *total* of the absolute values but the *difference* between them.[5] The addition of a negative and a positive magnitude does not produce an increase in amount but some sort of cancellation. Thus, if negative numbers can be understood in everyday situations by reference to debts, their addition to positive numbers must reflect this understanding.

3. *The minus sign as a representation of inversion.* The minus sign does not stand only for an operation or for the marking of a magnitude as negative. It has yet a third meaning that cannot be reduced to the first two: It stands also for inversion, as in $6 - (+4)$ or $6 - (-4)$. In these examples, " $-$ " is neither the sign for the operation nor a sign for the number but an indication that the operation must be inverted. Inversion is not encountered only in mathematics lessons where strings of signs like $6 - (-4)$ may be used. It is also encountered when people solving problems in everyday life attempt to reconstruct the starting point in a situation from the knowledge of its end-point. When a sequence of actions takes place in time, the end-point physically results from the interaction between the starting point and the actions that intervened. Given the end-point and the intervening actions, one can *mentally reconstruct the starting point by inverting the operations*. This ability to coordinate inverse operations and relations in a system is what Piaget referred to as "reversibility." Inversion applies to operations and relations but not to magnitudes (Piaget & Garcia, 1971).

Two questions are worth raising about the relationship between these different meanings of the minus sign. First, how do children coordinate these three different meanings? Must they be constructed together, does one have precedence over the other, or are they relatively independent? Second, how does the connection between one of the meanings and the written sign affect the solution of problems in which other meanings are involved. This question is particularly interesting because there is no distinction in the written mathematical representation of the different meanings: The same sign is used in any of the situations.

Two Empirical Studies Involving Negative Numbers

The experiments reported in this section analyze some of the questions raised earlier. More specifically, three questions will be investigated. First, we will look for evidence regarding the development of the understanding of negatively marked magnitudes and inversion outside school. There is already much evidence showing that some meanings for the operation of subtraction develop independently of schooling (see, e.g., Carraher, 1988), but little has been done on the two types of meaning investigated here. Second, we will try to see how these everyday conceptions are influenced by the use of the written minus sign, initially learned in school to represent the operation of subtraction. Finally, we will analyze the difficulties posed by inversion in both oral and written problem solving.

The negatively marked magnitudes investigated here were debts and losses after a transaction. A debt may be measured as an amount of money, but marked as different from income. Marking a debt as negative when solving a problem does not mean using written signs; it only means that one knows what other types of money to add it to. Expenses and debts are added together on one hand, whereas different amount of income obtained through consecutive sales are added separately. The conventional signs + and − may still be reserved by subjects for operations.

In order to investigate whether this type of meaning of negative numbers develops in everyday life outside school, it is necessary to interview subjects without instruction on negative numbers about tasks that relate to everyday life situations. Most of the studies about the understanding of negative numbers described in the literature seem to examine the performance of already instructed subjects (e.g., Bell, 1980; Janvier, 1985) or look at subjects working in situations that are not typical of everyday life, such as tasks involving movements on a number line (Galbraith, 1974). These studies have produced interesting results but cannot clarify the everyday conception of negative numbers. An exception to this rule is found in the work by Santos (1990), who investigated the understanding of debts and losses among unschooled and semischooled subsistence farmers. Despite their lack of instruction, farmers were quite capable of understanding and dealing with debts and losses, and of handling well addition of directed numbers in this context.[6] Santos observed that her subjects were equally successful in working with input data (such as production expenses and income from sale) and with relations (profit or loss when expenses and income had been taken into account). They were also able to take notes of losses and profits as the problem was presented; they spontaneously organized these in separate columns and carried out the computations at the end. Signs were neither used to mark integers as positive or negative nor used to indicate operations. This finding could be explained by the fact that subjects had very low levels of instruction and had not 'earned about signed numbers. Santos's study leaves little room for doubt about the existence of an intuitive conception of negative numbers as magnitudes developed independently of schooling. However, it leaves unanswered questions about the relationship between the different meanings of negative numbers and their interaction with signed written representation.

Study 1

This first study aimed at extending Santos's findings to schooled populations, investigating further the understanding of negative numbers in the context of problems about profits and losses, and analyzing the effect of schooling and written representation on problem solving.

In order to separate out the effect of schooling from that of age, two types of contrast were used. The first was a contrast between the performance of children and that of adults enrolled in evening classes at the same grade level as the children. It is possible to find such populations in Recife, Brazil, where this study was carried out, as a consequence of patterns of school attendance among working class families. State schools in Recife cater mostly to poor children, who may enter school later than the expected initial age of 7 and show high dropout and failure rates. Thus the age range for any grade level is high. When youngsters reach age levels considered out of range for the grade they are at, they are transferred into evening classes, which may also be attended by adults who dropped out and returned to school at a later point. The minimum age for evening classes is 18. Significant differences between younger and older pupils in the same grade level in this study would speak in favor of an age effect rather than a general effect of schooling; significant differences between grade levels irrespective of age would speak in favor of an effect of schooling rather than age.

The second contrast was between the performance of subjects with different levels of instruction: before, immediately after, and one year after instruction on signed numbers (respectively, fourth-, fifth- and sixth-grade instruction having been provided in the second semester of fifth grade). Both children and adults were involved in these comparisons.

To investigate the effect of using written versus oral representation during problem solving, we either asked subjects to solve the problems without making paper and pencil available during the experimental session or gave them paper and pencil and asked them to write down the information and solve the problems in written form.

Subjects

We interviewed 72 subjects (36 youngsters in the age range 10 to 17 years and 36 adults in the age range 18 to 23) who were enrolled in grades 4 ($n = 24$), 5 ($n = 24$), or 6 ($n = 24$) in state schools in Recife, Brazil.

Procedure

Subjects were seen individually and asked to solve the following problems: (1) 30 problems related to debts and losses in hypothetical situations about agriculture, and (2) 10 problems about movements on a number line. This second group of problems will not be analyzed here. Half of the subjects were asked to solve the problems orally, and the other half solved the problems in writing. This condition was orthogonal to the order of the tasks, which was AB for half the subjects and BA for the other half. Subjects were allowed to use whatever written representation they wanted as long as they wrote down the information. After each response, subjects in both groups were asked to explain how the answer was obtained.

All problems involved a starting point (how much money there was to begin with and an end-point (the final financial situation), as well as two intermediary transformations, which could involve either profit or loss. Problems were divided into five groups according to their structure. Group 1 had zero as the starting point. Group 2 problems had a positive amount as the starting point. Group 3 problems had a debt as the starting point. In groups 1 through 3, the starting point was known and the end-point was to be calculated. Thus, problems in these groups did not require inversion but involved only the meaning of marked magnitudes. Groups 4 and 5 included problems with end-points given and starting points to be calculated; thus, they involved inversion. In group 4 the end-point was positive, and in group 5 the end-point was a debt. Each group had two problems with transformations in the same direction (two profits or two losses) and four problems in which the directions of the transformations were different, two with profits larger than losses and two with losses larger than profits. Problems were presented in a fixed order (from easier to more difficult according to pilot data collected in the same population). Two sample problems are presented in Figures 5-2 and 5-3 in the next section.

Responses in both the oral and the written condition were considered correct if they were numerically correct and the subjects correctly indicated (either orally or in writing) whether they thought the result to be a profit or a loss.

Results and Discussion

We initially analyzed the results to evaluate whether having solved the number line task influenced performance in this task. No effect was obtained (in an analysis of variance, or ANOVA, with order of tasks and grade level as independent variables), and all results were then analyzed together, in disregard of the fact that some subjects had solved another task about negative numbers before working on this task. We then looked at the influence of the three main variables on performance: (1) age group (10–17 versus 18 and above); (2) grade level (fourth, fifth, or sixth grade); and (3) condition of problem solving (oral versus written). Figure 5-1 presents the mean number of correct responses for each grade level by condition. An analysis of variance indicated that neither age nor grade level had a significant effect on performance. In contrast, subjects in the oral condition did significantly better than those in the written condition (F obtained for condition = 14.32; $p < .001$).

FIGURE 5-1

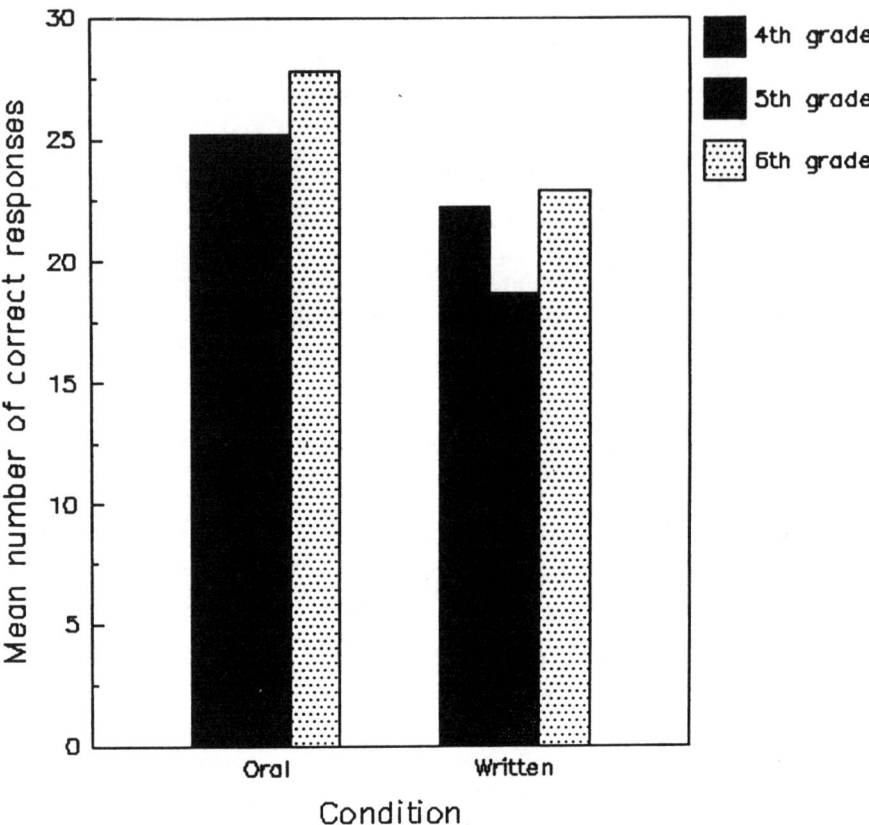

The absence of a reliable effect of grade level is surprising; despite the fact that fifth- and sixth-graders were receiving instruction on signed numbers, they did not perform consistently better than those in grades 4. These results can be interpreted as an indication that subjects were solving the problems using concepts and strategies learned outside of school and that school instruction was not effective in changing their conceptions. This interpretation is reinforced when one considers that school arithmetic is mostly written in Brazil, whereas out-of-school arithmetic is oral (see Carraher, Carraher, & Schliemann, 1987) and that writing down the information interfered with rather than facilitated problem solving.

Further analyses of the protocols revealed the reasons for the negative interference of written procedures with correct solutions. When asked to solve the problems on paper, subjects did not use the conventional signed number representation for either the initial amount or the final amount (only 2 of the 72 subjects did so). They often used letters to mark whether the initial amounts were debts or assets. The symbols + and − were reserved for operations whenever used. Subjects working in the written condition also tended to make the temporal order in the problem correspond to the order of computation and faced difficulty when this was not possible. This mixture of intuitive and formal representation interfered with solution. Figures 5-2 and 5-3 present sample problems and subjects' attempts at written solutions.

Both subjects faced difficulty with the written problem. J. C. wrote "+10" to indicate the increase in amount and then interpreted the sign as an indication to add up the numbers despite her knowledge, displayed later in the oral solution, that the initial debt and the gains should not be added up. U.S. solved the problem initially without writing anything down and did so correctly, adding the losses and working backwards to the initial situation. When he wrote down the information, he was able to add the losses but did not know how to perform the inversion in a formal way. He produced an operation (−20) not mentioned in the problem which would simply "fill the blank," since the didactic contract (see Chevallard, 1985) in mathematics classes requires that all numbers be connected by operations. An adequate formalization, [10 − (−30)], could not be produced despite the subject's ability to carry out inversion in an intuitive fashion.

FIGURE 5-2

Problem (from group 3): *Seu Pedro* (the farmer's name) started the season with a debt of 10 *cruzados* (Brazilian currency). He planted manioc and beans. He gained 10 on the manioc and 20 on the beans. What was his situation at the end of the season?

    J.C. (19 years old, sixth-grader, assigned to the written condition) wrote line 1 without indicating, and line 3 without a sign. Added all the numbers, obtaining 40, which was written on line 4.

    J.C.: Profit. No, it's not that. I can't do it on paper.

    I: Why?

    J.C.: He had a profit of 10, paid the 10 he owed. Then he still has his profit from the beans, he has 20.

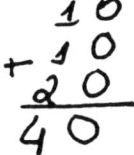

FIGURE 5-3

Problem (from group 4): *Seu Severino* (the farmer's name) ended the season with 10 *cruzados*. He had planted sugar cane and corn. He lost 10 on the sugar cane and 20 on the corn. What was his situation when he started out the season?

U.S.: (22 years old, sixth-grader, assigned to the written condition, answered without having written down anything): He had 40 *cruzados* to begin with.

Interviewer: Was that a debt or a profit?

U.S.: Not a debit, profit.

I: Can you do it in writing? You are doing it so quickly, I wish you could do it in writing so I could see how you do it. U.S. writes down lines 1 and 2 below indicating that the amounts are debts *(prejuizo)* through the letter P. He then adds the two debts (line 3). This is followed by the arrow and the notation of 10 L *(lucro)*, which is the end result. After a pause, he writes on line 4 " − 20," an amount that allows him to end up with the number 10 (line 5). Asked to explain this procedure, he said: No, I can 't do it on paper.

I: Do you know why?

U.S.: It is not 20, he started out with 40. If he lost 30 and ended up with 10, he started with 40.

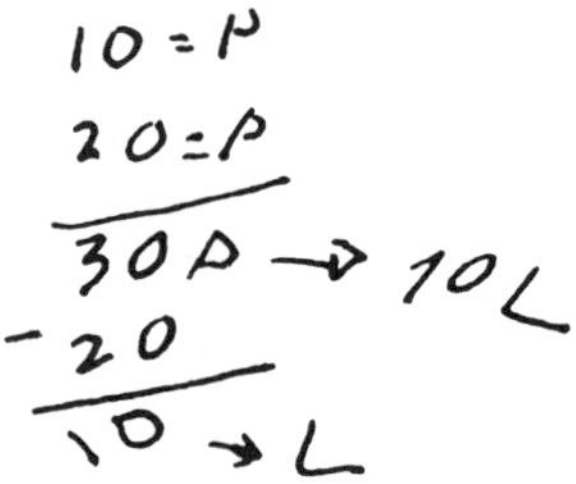

$$10 = P$$
$$20 = P$$
$$\overline{30 P} \rightarrow 10 L$$
$$-20$$
$$\overline{10} \rightarrow L$$

Further quantitative analysis was carried out to investigate the effect of problem type on performance. This analysis showed no significant main effect of grade level and no significant interaction between problem type and grade level. However, significant effects were obtained for condition (oral versus written, $F = 14.96; p < .001$), problem type (groups of problems 1 through 5, $F = 25.64, p < .001$) and for the interaction between problem type and condition ($F = 2.59; p = .037$). The different types of problem had been previously organized by order of difficulty, and this order was not replicated in only one (out of four) steps for the oral condition and was replicated for all the steps in the written condition. Statistical comparisons between the mean number of correct responses for the different types of items were carried out within both the oral and the written conditions. Significant differences were obtained (using the Newman-Keuls method) only when direct and inverse problems were compared, but the difference between the most difficult type of direct problem and the easiest type of inverse problem was not significant in either the oral or the written condition. Figure 5-4 presents the mean number of correct responses as a function of problem type for each condition given the significant interaction between problem type and condition.

This study leads to some clear conclusions and also to some hypotheses. First, it is clear that negative numbers are not constructed only in mathematics classes through formalizations. Although the inverse problems were more

FIGURE 5-4

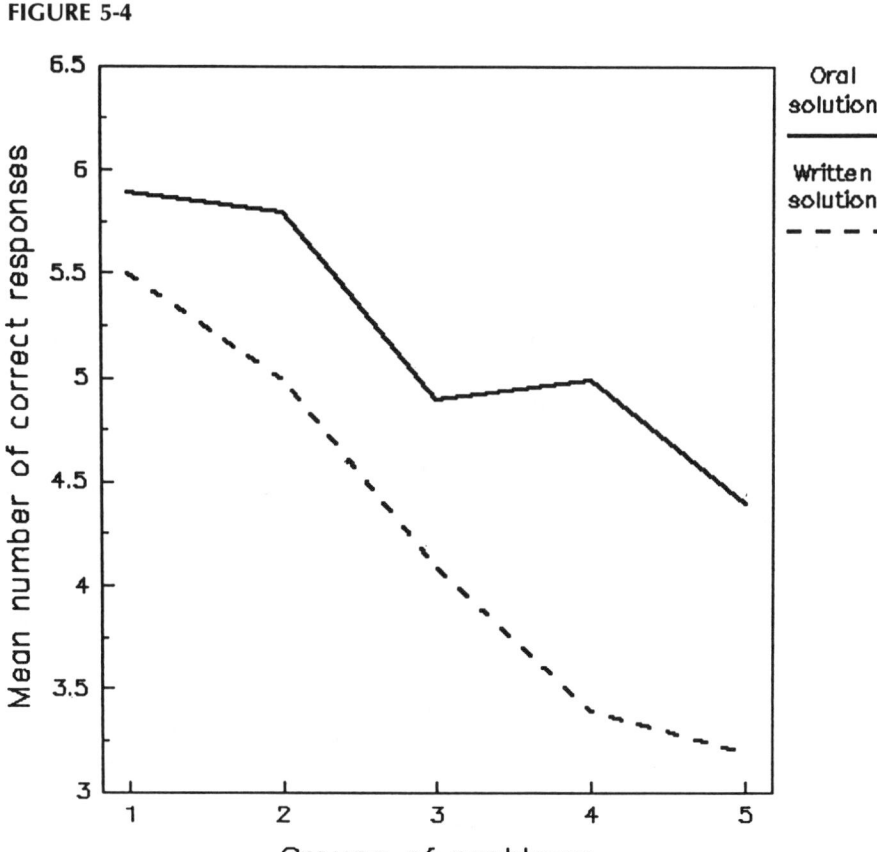

difficult to solve than the direct ones, there was evidence that uninstructed subjects understood all three meanings of the minus sign. Subjects who had never received instruction on negative numbers in both age groups clearly demonstrated an ability to solve loss and profit problems; they showed 84% correct responses in the overall task when it was presented orally and 74% correct responses when they were required to write the numbers before responding. Second, teaching did not appear to be effective in improving problem-solving strategies. Instructed subjects did not do significantly better than uninstructed subjects. Moreover, the use of written representation, which is typical of school instruction, did not reflect the adoption of the conventional representation of signed numbers and interfered with, rather than improved, performance. Qualitative analyses of the protocols suggested the hypothesis that this interference might result from the attribution of only one meaning to the minus sign—that of subtraction. This hypothesis was then tested in a second study in which we tried to counteract the negative effects of the written condition.

Study 2

This second study was carried out in order to test whether the introduction of written representation of signed numbers could be done without disrupting performance in problem solving by making it clear to the subjects that the signs before the numbers indicated the type of magnitude rather than the operation. In the previous study we had seen that subjects sometimes used letters to mark the initial amounts of money mentioned in the problems as either debts or assets and reserved the signs + and − to denote operations. It was expected that, if we succeeded in leading subjects to consider the signed number representations as markings for the type of quantity, they would do just as well in the written as in the oral condition and would do significantly better in the written condition than subjects in the previous study, whose performance had been disrupted by the written representation, but similarly to those solving problems in the oral condition.

Subjects

Forty-eight adults enrolled in evening classes in the age range 21 to 25 years (mean = 22.3) were interviewed. As the study was carried out in the first half of the academic year, subjects in grade 5 had not yet received instruction on signed numbers. In order to interview subjects in three grade levels (corresponding to before, just after and sometime after instruction on signed numbers), subjects were sampled from grades 5 ($n = 16$), 6 ($n = 16$), and 7 ($n = 16$).

Procedure

Each subject was seen individually on two separate occasions. In the first interview, subjects were presented with the same set of problems described in study 1 and asked to solve the problems orally. About one or two weeks later, the second set of problems was presented. Subjects were told: (1) that a farmer had written down his losses and profits as shown on the cards presented to them; (2) that he had used a minus sign before a loss or a debt and a plus sign before a credit or a profit; and (3) that the pieces of information on the card were about his initial situation, the transformations thereafter, and the final result, each being always placed in a particular position on the card; (4) that some of the information was missing, and the subjects' task was to complete the farmers' notes by putting in that information. As in the previous study, in problem groups 1 through 3 the missing information was the final situation of the farmer, whereas in problem groups 4 and 5 the initial situation was missing and inversion was required.

Results and Discussion

The effect of condition of problem solving (oral versus written) and grade level on performance was statistically investigated through ANOVAs, which were carried out both for the groups of problems separately and for the total score in the task. No significant effect of either condition or grade level was observed, a result that demonstrated that the written condition in this case did not disrupt performance. Figure 5-5 presents the mean number of correct responses for each grade level and each condition.

The performance of the subjects in this study was contrasted to that of subjects from study 1 because we expected no difference across studies in the oral condition but a clear difference in the written condition. Two groups of subjects were formed, one including those subjects who had not received instruction on signed numbers (grades 4 and 5, respectively, for studies 1 and 2) and one of those who had received instruction (grades 5 and 6 for study 1 and grades 6 and 7 for study 2). Despite the differences in grade level, the comparison is still considered appropriate, for two reasons. First, the actual difference in schooling between subjects in grade 4 in study 1 and those in grade 5 in study 2 was of only one term, because grade 4 subjects were tested in the second half of the year and grade 5 subjects were tested in the first half. Second, in neither study did grade level emerge as a significant effect. Figure 5-6 does not show a clear difference between subjects in study 1 and study 2 in the oral condition. A noticeable difference appears only in the written condition, with subjects in study 2 showing better performance in the task. No statistical comparisons were considered appropriate in this case, but the pattern of results already described (significant differences between oral and written conditions in the first study but not in the second) seems quite clear.

FIGURE 5-5

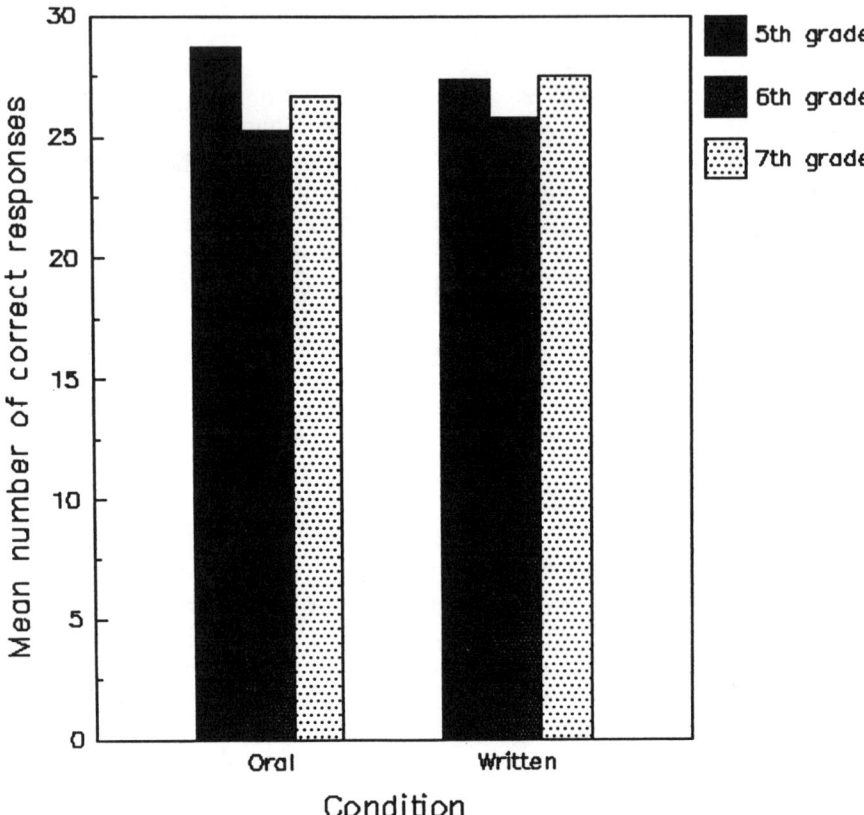

These results support the hypothesis that performance in the written task in study 1 was disrupted by the subjects' attribution of only an addition/subtraction meaning to the plus and minus signs. This interpretation interfered with their use of the notion of quantities marked either negatively or positively when debts had to be added and profits subtracted. A maneuver that made the meaning of the signs clearly connected to the marking of quantities was successful in maintaining performance at the same level in the written as in the oral condition.

Conclusions

These studies support the idea that a basic understanding of signed numbers can be developed in everyday life. Although all three types of meaning may

FIGURE 5-6

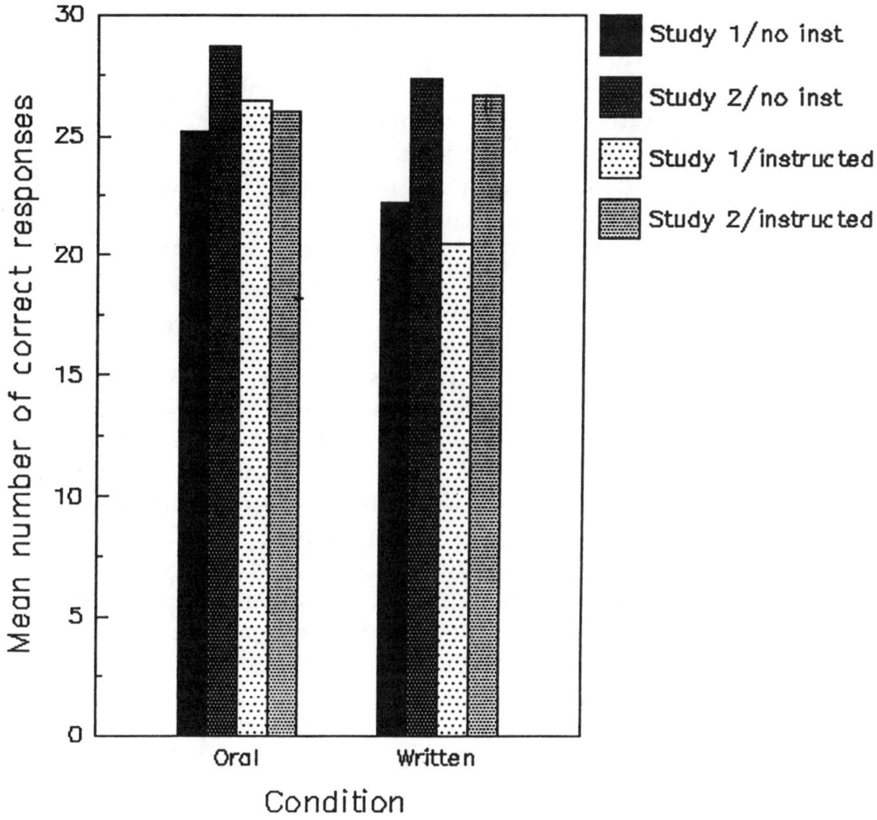

be encountered outside school, the operational and the marked magnitudes meanings are better understood than is inversion. These results confront the notion that signed numbers are simply formalizations that must be constructed in mathematics classes through formal demonstration. The availability of semantic representations for signed numbers constructed in everyday life may, at first glance, contrast with pupils' difficulties with this same topic in the classroom. Gallardo and Rojano (1987), for example, reported great difficulties among pupils in understanding and operating with negative numbers. This difficulty, according to our results, may not be conceptual in nature but may result from a notational confusion. Because the same notation is used to refer to different semantic situations, pupils may lose perspective on how to proceed in even simple problems.

The need to distinguish between the availability of a semantic representation for negative numbers and the use of the mathematical representation through signs was demonstrated in two ways in the studies reported here. First, in study 1 subjects working out problems without written representation did significantly better than those who had to use written numbers. Second, study 2 showed that solving problems in which numbers are presented in writing does not become any more difficult if notational confusion is avoided. Taken together, these results indicate that what has often been viewed as a conceptual difficulty may, in fact, be a difficulty introduced by the notational conventions we use. Similar results were observed by Bell (1980) in a study about movements on the number line, in which it was found that children became more confused if they tried to record the problem before calculating.

Mathematics teachers can thus rely more on children's everyday concepts in teaching signed number and concentrate on the difficulties of learning conventional notation. These difficulties, however, should not be viewed as lesser concerns in mathematics education. Two arguments are worth considering. First, notations are often important tools in mathematics, and they must be learned if pupils are going to live well in a mathematical culture. Place value notation, for example, poses problems peculiar to the notation itself and independent of the understanding of number, such as the use of zero as a place holder. These problems are well worth surmounting in view of the power of this notation in making computations easier and in coordinating integers and decimal fractions. Similarly, signed-number notation may cause pupils some problems, but the reasoning power it purchases in algebra makes its learning worth the effort. Second, the use of the same notation for three different referential situations is not arbitrary and accidental as is, for example, the use of the same word, *table* to signify both a piece of furniture and a particular way of organizing data. The same notation is used in mathematics because the three situations, though distinct from the semantic viewpoint, obey the same rule with respect to operations.

Mathematics educators have been at least to some extent aware of the fact that some problems with signed numbers are notational rather than conceptual. In fact, some teaching schemes (see Galbraith, 1974) propose that the sign for the operation and the sign for the number be distinguished in the initial phases of teaching directed numbers. It is, however, necessary to evaluate whether or not the use of two notations is a solution to the problems raised earlier. Two notations might actually compete with the goal of bringing pupils to the realization that the different semantic interpretations are represented by the same signs because they obey the same rules.

The studies presented here suggest a new direction. Teachers may want to strengthen the understanding of each meaning independently of notation and then bring the same notation into each field separately before pupils become fully aware of their convergence.

Notes

1. *Editors' note*: This may not be historically correct. See Chapter 4.
2. It is possible to distinguish other meanings for the minus sign. Vergnaud (1982), for example, distinguishes between states, relations, and transformations. A comparison of models is, however, beyond the scope of this chapter.
3. *Editors note*: The "take away" meaning of 5 − 2 inevitably presents difficulties in making drawings. To see the problem, compare 5 − 2 with 5 + 2, where the problem does *not* occur. For 5 + 2 we can draw five objects, and then also draw two more. Counting them all, one is seeing 5 + 2, as desired. But contrast this with 5 − 2. If we draw five objects (which would be correct), then we must not draw any more. According to the mathematical notation 5 − 2, *five is all there are*. We *Must* not draw more!

    How, then, can we represent 5 − 2 pictorially? Not by drawing two additional objects. Elementary textbooks sometimes try to show this by, perhaps, picturing five birds sitting on a tree limb and then showing two of them flying away while three remain on the tree limb. Another device is to show five cans of tomato soup on a grocer's shelf and to have a customer appear with a shopping list that calls for the purchase of two cans of tomato soup. After this transaction, what the grocer will have left will be 5 − 2 cans of tomato soup.
4. *Editors' note*: Please refer to Chapter 4 for the distinction between the "two-attribute" conceptualization of positive and negative numbers, as opposed to the more sophisticated "one-attribute" concept.
5. *Editors' note*: This, of course, is a definition of the "two-attribute" conceptualization of negative numbers. It is only when the combining process is seen as *addition* (rather than subtraction) that the more powerful "single attribute" conceptualization has been achieved, so that $x + y$ can *always* be considered to be addition, no matter what signs x and y may have. See, for example Davis (1984, pp. 165–170).
6. *Editors' note*: Here, too, we are *not* dealing with the powerful single-attribute meaning of signed numbers but, rather, with an important experiential precursor of that concept.

References

Bell, A. (1980). Developmental studies in the additive composition of numbers. *Recherches en Didactique des Mathematiques*, 1, 113–141.

Brown, M. (1981). Number operations. In K. M. Hart (Ed.), *Children's understanding of mathematics: 11–16* (pp. 23–47). London: Alden Press.

Carpenter, T., & Moser, M. (1982). The development of addition and subtraction problem solving skills. In T. Carpenter, M. Moser, & J. Romberg (Eds.), *Addition and subtraction: A cognitive perspective* (pp. 9–24). Hillsdale, N J: Lawrence Erlbaum.

Carraher, T. N. (1988, April). *Adult mathematical skills: The contribution of schooling.* Paper presented at the annual meeting of the American Educational Research Association, New Orleans.

Carraher, T. N., & Bryant, P. E. (1987, July). *Addition and subtraction in everyday life and arithmetic.* Paper presented at the China Satellite Conference of the International Society for the Study of Behaviour Development, Beijing.

Carraher, T. N., Carraher, D. W., & Schliemann, A. D. (1987). Written and oral mathematics. *Journal for Research in Mathematics Education*, *18*(2), 83-97.

Chevallard, Y. (1985). *La transposition didactique: Du savoir savant ou savoir enseigne*. Grenoble: La Pensee Sauvage.

Davis, R. B. (1984). *Lerarning mathematics: The cognitive science approach to mathematics*. Norwood, NJ: Ablex.

Galbraith, M. J. (1974). Negative numbers. *International Journal of Mathematics, Education, Science and Technology*, *5*, 83-90.

Gallardo, A., & Rojano, T. (1987). Common difficulties in the learning of algebra among children displaying low and medium pre-algebraic proficiency levels. In J. Bergeron, N. Herscovics, & C. Kieran (Eds.). *Proceedings of the Eleventh International Conference on Psychology of Mathematics Education* (pp. 301-307). Montreal.

Hughes, M. (1986). *Children and number: Difficulties in learning mathematics*. Oxford: Basil Blackwell.

Janvier, C. (1985). Comparison of models aimed at teaching signed integers. In L. Streefland (Ed.), *Proceedings of the Ninth International Conference for the Psychology of Mathematics Education* (pp. 135-140). Utrecht: University of Utrecht.

Piaget, J., & Garcia, R. (1971). *Les explications causales*. Paris: Presses Universitaires de France.

Riley, M. S., Greeno, J. G., & Heller, J. I. (1983). Development of children's problem-solving ability in arithmetic. In H. P. Ginsburg (Ed.), *The development of mathematical thinking* (pp. 153-196). New York: Academic Press.

Santos, A. M. (1990). *Compreensao e uso de numeros relativos na agricultura e na escola*. Master's thesis, Mestrado em Psicologia, Universidade Federal de Pernambuco, Recife, Brazil.

Vergnaud, G. A. (1982). A classification of cognitive tasks and operations involved in addition and subtraction problems. In T. Carpenter, M. Moser, & J. Romberg (Eds.), *Addition and subtraction: A cognitive perspective* (pp. 39-59). Hillsdale, NJ: Lawrence Erlbaum.

CHAPTER SIX

Understanding Mathematics
The Impact of Technology

ANTHONY RALSTON

Abstract

The purpose of this chapter is to show how technology is having an impact on our perceptions of what mathematics is more or less important than previously and on how technological changes should be affecting the teaching of mathematics. To this end we discuss the increasing importance of mathematics in many professions that have been wrought by computer technology and consider whether this implies a need for revolution or evolution in what mathematics is taught and how. Then we discuss the notion of a zero-based curriculum in which all topics must be justified by their current importance and the effect of technology on such a zero-based curriculum. Next, the impact of technology on the teacher of mathematics is considered with respect to the need to learn new material and new ways of teaching — indeed, to revamp in large measure what is now standard practice. The relation of research in mathematics education and technology is discussed from the perspective of a need for more symbiosis than there is now. Then the impact of technology on a drill-and-practice regimen in mathematics teaching is considered from the perspective of why it requires a rethinking of the "knowing is doing" viewpoint on learning mathematics. Finally, we make some remarks on when technology is appropriately used in teaching mathematics and when not.

I am indebted to a number of people who made comments on an earlier draft of this chapter. It is a pleasure to give particular thanks to Hugh Burkhardt, Doug Clements, Ed Dubinsky, Jim Fey, and Steve Maurer.

Throughout its history, until very recently, mathematics has been a *theoretical* discipline in the sense that *experimentation*, as generally understood, played a rather small role. True, mathematicians have often gained insight into their problems by doing some (arithmetic or symbolic) calculation or by working out special cases. But this was seldom a major portion of any mathematician's activity (barring some calculating prodigies like Gauss). It has been quite rare for extensive mathematical experiments to be performed for the express purpose of developing conjectures and theories. Computers have wrought a major change. Although it is still true that mathematicians trained before, say, 1970 use computers relatively rarely as other than calculational engines, an increasing number of younger mathematicians are learning to use the computer in its more significant role as a mind expander.

By contrast, the physical sciences and biology from the Middle Ages on all had an experimental component before they had any significant theoretical component. In these sciences, experiment has always been seen as the necessary underpinning of any theory. Experiments are, of course, used to test theories, but it is rare in these sciences for theories to be developed from whole cloth without the suggestiveness of experimental data on which to build them. Surely Lavoisier would never have propounded a theory of combustion if he had not performed many experiments first.

Another way of viewing the changes wrought on mathematics by technology is this: In addition to the two classical paradigms of science—theory and experiment—there is now a third—computing—which, through simulation using a mathematical model, is another way of getting a view of the world. Indeed, we may say that computing is to mathematics as laboratory experimentation is to the classical sciences.

Since experimentation inevitably involves some form of *technology*, even if no more than Archimedes' lever, it is hardly surprising that education in the physical sciences and biology has for many decades, perhaps centuries, used technology as a key ingredient in the teaching of these subjects both through laboratories but also—and more importantly for my purpose here—as an integral part of almost all lectures.

Mathematicians, on the other hand, have been very, very slow to adopt the use of technology as part of their teaching arsenal. Chalk and chalkboard are still the technology of choice for most mathematicians, with the more adventurous ones opting for overhead projector and screen. The inertia of history is a crucial reason for this. So, too, it must be admitted, is the fact that adoption of the only significant technology available for mathematics teaching, namely calculators and computers, involves, at least initially, much harder work than most mathematicians—college and university mathematicians anyhow—are willing to put into their teaching, a matter we return to later.

So we ask in this chapter: Can mathematics be properly taught in the last decade of the twentieth century without recourse to the use of technology in the classroom? Surely this is not a trivial question. The mere existence of technology is no excuse for revamping practice if that practice is currently satisfactory or, even if it is not, if there is no valid reason to believe that the technology will improve whatever is wrong. An attempt to answer this question will inevitably lead us to ask related questions about the mathematics curriculum and about research in mathematics education.

Is Technology More Relevant to Mathematical Education Than Heretofore?

Prior to, say, World War II, only a very small elite needed to know any mathematics beyond arithmetic and, perhaps, a soupçon of geometry in order to live their lives and to be able to perform their jobs. Moreover, that mathematics which was important, namely arithmetic, was needed by almost everybody in the developed world, at least. Addition and subtraction were useful almost every day to most adults. Since World War II two things have happened to change this picture. First, the rapid advancement of technology in virtually all spheres of life has created a much wider need for mathematical knowledge than heretofore; for example, both lawyers and doctors need increasingly to use technology, a relatively new development for both professions; if they are going to use it effectively and not just accept the results uncritically, they will need to understand considerable amounts about the mathematics that lies behind the screen displays and printed outputs; more generally, it is almost certain that the fraction of people for whom nontrivial mathematical knowledge will be important will continue to grow for some years.

Second, the value of hand arithmetic skill in and of itself has decreased rapidly and is now or soon will be negligible. The question remains of whether or not such skill is nevertheless necessary for the pursuit of higher mathematical knowledge. The essential point, however, is that now this question must be asked whereas previously it didn't matter much because the skill itself was necessary for everyday living.

These two factors together imply at least the possibility that a revolution is necessary in mathematics education in terms of both what is taught and how it is taught. To a considerable degree mathematicians and mathematics educators have been unwilling to face this possibility. Under the guise of a what has worked in the past will work in the future attitude and despite mounting evidence that, whether or not current approaches ever worked very well, they don't work well enough now, many research mathematicians and a considerable, though dwindling, number of mathematics educators have been content with a business-as-usual or, at best, a nibbling-at-the-edges approach. But it won't compute! Mathematics education in the United States, at least, is failing and failing increasingly badly. Sadly, the reasons it is failing go far beyond issues of curriculum and pedagogy. Still, even if the societal reasons for the parlous state of mathematics education today were somehow miraculously solved, things would not improve much without major changes in curriculum and pedagogy. Mathematics educators themselves cannot do much about the impact of television or parental disinterest in their children's education or the breakup of the traditional family, but they can at least address themselves to how mathematical education should be pursued if none of these disasters existed. To do so inevitably requires that the impact of technology—computer technology—be reckoned with in all aspects of mathematics education.

A Zero-Based Mathematics Curriculum

Suppose Archimedes and Euclid had had calculators and computers. How would mathematics education today differ from the present reality? Who can doubt that it would be vastly different, if not in underlying subject matter, then at least in approach and pedagogy? And few would argue that the result would have been less mathematical progress than we have seen because children would not have been drilled to become fluent in pencil-and-paper arithmetic.

Or ask this: Suppose there were no such thing as mathematics education today and we had to design it from scratch. What would it look like? Who doubts that it would look quite different from what we have today?

The point of both these gedanken experiments is to illustrate that the system we have today is not God-given, or the best of all possible worlds, but, rather, one that has developed over centuries in response to the prevailing realities at the time. Evolutionary development rather than revolution is the normal path in the development of intellectual disciplines. But from time to time in all disciplines there are what Thomas Kuhn (1970) calls paradigm shifts (the theory of relativity in physics was one such shift), which result in discontinuities of perception and belief. A crucial question in mathematics education today is whether the impact of calculators and computers on mathematics education creates a paradigm shift that should force major changes in what we do. The answer is certainly not obvious because it can be argued that, despite various discoveries that resulted in paradigm shifts in the history of mathematics (e.g., the invention of the calculus, Godel's theorem), mathematics education has had an entirely evolutionary development.

There is a considerable body of opinion (supported, it might be added, by very few facts), particularly prevalent among mathematics researchers, that our current failures in mathematics education have nothing to do with curriculum or pedagogy but are instead the result of failures to do a good enough job teaching the traditional curriculum in the traditional manner. It is hard to refute this opinion for at least three reasons:

- There are clearly deep-seated societal factors, some of them mentioned previously, that would have resulted in serious deficiencies in mathematics education even if we were teaching the perfect curriculum with superb teachers.
- If the present curriculum — or any curriculum — were well enough taught, then mathematics education problems would be at most modestly serious and quite tractable.
- There has been no demonstration, much less a proof, that significant, perhaps radical changes would result in any improvement.

It is, however, as clear as anything in education is that the quality of mathematics teaching in the United States, at least, will not be good enough in any foreseeable future to make the curriculum almost irrelevant. In fact, the contrary is almost certainly the case: A declining quality of mathematics teaching, which it would be kind to call pedestrian, means that increasing numbers of students are sure to be turned off to mathematics by the current curriculum.

Not to despair. There is evidence that the use of technology in teaching mathematics is not only not harmful but is positively helpful (Hembree & Dessart, 1986; Clements & Nastasi, 1992). True, the evidence is far from overwhelming, not least because controlled studies are so difficult to perform and because the Hawthorne effect is so common to much of the research that is done. Still, although much of the research is only moderately compelling, at least it gives no cause to believe that we would do children permanent damage if they were taught a very different mathematics curriculum than is usual in American schools and those of many other countries.

What might such a curriculum look like? Well, the second of the experiments alluded to earlier suggests the right approach to answering this question. Although assuredly there *is* a mathematics curriculum and even the most wild-eyed revolutionary would be forced to admit that somehow a path must be found from here (where we are) to there (the new curriculum), it is nevertheless instructive and practical to consider a *zero-based mathematics curriculum*. By this I mean only that the mathematics curriculum in K–12 (but why not the college curriculum, too?) should be viewed from the perspective that nothing is sacred, that each topic in each year must be justified on its present merits and not just because it is there and may have been there for centuries.

In one sense, the result of developing a zero-based curriculum would not be radically different from the current mathematics curriculum. There would still be lots of arithmetic, algebra and geometry. But in another sense the changes would be quite radical. Considerable portions of the current K-12 subject matter would just disappear, to be replaced by topics that currently are not covered at all or get much too short shrift. The motivations and justifications for most of these curricular additions and subtractions would be technological.

Any topic to be included in a zero-based mathematics curriculum should satisfy at least two of the following criteria:

1. It should be of intrinsic mathematical importance. Certainly large portions of arithmetic, algebra, geometry, and analysis fall under this rubric.
2. It should be of significant mathematical interest. That is, a topic may be included if, though not of major importance, it can add interest to the study of mathematics or motivate students to study mathematics. New mathematics with contemporary applications and various topics in recreational mathematics, such as the game of Nim, would fall in this category.
3. It is important for the study of further mathematics. Thus, the algorithms of pencil-and-paper arithmetic may no longer have any intrinsic importance but, if learning them is necessary for the later study of mathematics, they should be included.
4. It has practical value that will be useful to almost all who study it. Examples would be percentages and various aspects of financial mathematics; school mathematics must not be independent of real mathematics.

Additionally, of course, the intellectual level should be appropriate for the grade in which the topic is to be taught. The foregoing list purposely avoids issues of review and repetition. Suffice it to say that a zero-based curriculum could easily include the spiral idea but, hopefully, not include it, as too often seems to happen, in its form of repetition without growth.

Technology would affect the development of a zero-based curriculum in two ways:

1. The importance of topics in mathematics, as well as their intrinsic interest, has been profoundly affected by the existence of computers and calculators. With arithmetic this is obvious; for various portions of algebra and calculus, specifically those portions that deal primarily with symbol

manipulation (e.g., polynomial arithmetic, differentiation), it is almost as obvious. As computer capabilities in symbol manipulation, logic, and even proof are further developed, other portions of mathematics will lose their intrinsic importance.

The conclusion from this is not that such topics should be dropped from the curriculum. Learning them may still be necessary for the study of higher mathematics or for practical reasons. But the burden of proof (as opposed to gut feeling) should be on those who wish to include topics whose importance has been rendered problematic at best by computers. Certainly pencil-and-paper arithmetic is not necessary for the further study of mathematics (which is, of course, not to say that an understanding of the operators of arithmetic and facility in mental arithmetic and estimation are not as or more important than they have ever been).

2. A curriculum is more than a list of topics or a scope and sequence list. A proper curriculum should also specify (broadly rather than narrowly) the pedagogy most appropriate for each topic. In mathematics there are few topics at any level whose presentation cannot be enhanced by the use of computers or calculators in the classroom. There is considerable evidence to support this assertion at all levels of K–12 as well as for college mathematics (Hembree & Dessart, 1986; Clements & Nastasi, 1992). Still it must be admitted that, although a number of teachers and researchers have developed software and calculator exercises suitable for classroom use, the number of mathematics educators who use these tools is still very small, far too few to provide anything approaching a proof of their efficacy. Still I submit that, for example, it is intuitively clear that the dynamic capabilities of computer graphics must provide a better medium for teaching a mathematics of change (e.g., calculus) than the essentially static mechanisms of chalkboard or overhead projector.

Is discussing a zero-based curriculum just fantasy, given the tremendous inertia of the current curriculum? Perhaps. Mathematics education reformers generally seem unwilling to embrace such a radical notion. Not long ago I chaired a committee on the K–12 curriculum of the Mathematical Science Education Board, whose report led (loosely) to a National Research Council (NRC) report (National Research Council, 1990). The idea of a zero-based curriculum was prominent in the report of the committee but, after digestion by the NRC, did not see the light of day in the final report. Still I persist in thinking that the design of a zero-based curriculum would be, at least, a useful intellectual exercise which could be influential in the movement to reform mathematics eduction. I call your attention to Benezet's (1935, 1936) experience sixty years ago, which shows that radical reform, properly led, can be implemented and effective.

Technology and the Mathematics Teacher

The notion of a zero-based curriculum can be both exciting and threatening to the teacher of mathematics. The attraction is clear. There is an opportunity for innovation and creativity that few teachers of mathematics at any level now experience. But the threats are real and require some discussion.

The first would be the necessity for learning and teaching mathematics that will be unfamiliar to many current teachers of mathematics. At the elementary school level, it would no longer be sufficient for a teacher only to be a competent pencil-and-paper arithmetician. Not only would familiarity with calculators and computers be necessary, in addition, an appreciation of the patterns of arithmetic, which is essential for calculator usage, would become necessary. Beyond this, as the role of arithmetic in the elementary school curriculum declines, it will be both necessary and desirable to include other branches of mathematics in the elementary school curriculum. There is (in theory, at least) some geometry in elementary school now, but the role of this branch of mathematics needs to be considerably expanded in elementary school. There is interesting and useful geometry that can be taught in every elementary school grade. In addition, there are branches of mathematics that seldom are mentioned in elementary school today that need to be introduced because they are important and interesting and are also within the grasp of elementary school students. Examples are probability and data analysis, both of which satisfy all four criteria in the previous section.

Secondary school teachers also will have to expand their mathematical horizons as the secondary school curriculum becomes more than just algebra, geometry, trigonometry, and precalculus. This is already happening as some statistics and data analysis makes its way into the high school curriculum because of the influence of computing. But other mathematics whose importance is also increasing rapidly because of computers, such as combinatorics, will also become more important in the secondary curriculum. Indeed, numerous aspects of so-called discrete mathematics, all of them made much more important by computers, are strong candidates for introduction into the secondary curriculum.

But the most significant impact of technology on the mathematics teacher will probably be caused not by the changes in curriculum but by the changes in pedagogy that should be introduced. We have noted that there is no mathematics suitable for the secondary school student whose presentation cannot be enhanced by a computer in the classroom. This poses a major challenge for all teachers of mathematics. It is not just the hardware and software problems and the logistics of making classrooms user-friendly for teachers and students although these are real problems that can cause considerable frustration. More difficult is the need for teachers to prepare almost wholly anew for teaching with technology as well as the need to consider pedagogical approaches that calculator and computer technology make possible or even necessary. A computer is not just another classroom artifact like a chalkboard or an overhead projector. It is a new species whose effective (as opposed to pedestrian) use requires rethought preparation for almost every class with regard to both the subject matter itself and the interaction of teacher with students through small-group and individual instruction as well as traditional instruction to the entire class.

Where to find the human and financial resources to do what the technology will demand is going to be a major challenge for mathematics educators and the education establishment generally.

Understanding Mathematics: Technology and Research on Mathematics Education

For the moment, accept the assumption that a mathematics curriculum appropriate for the 1990s would be considerably different from the present one and would be one in which computers and/or calculators were tools in an almost daily use in the classroom. What would this imply about the thrust of current and past research in mathematics education? Far too little of current research in mathematics education is devoted to understanding the uses and limitations of technology for teaching. Historically, and still today, a great deal of the research in mathematics education has focused on fairly specific aspects of the current curriculum (e.g., the effects on learning of various borrowing strategies in subtraction). Almost all of this research is irrelevant to the current situation.

Nevertheless, it would be wrong to be too pessimistic about the role research in mathematics education can play in reforming the mathematics curriculum and mathematical pedagogy. There is a considerable and growing amount of research focused directly on the role of technology in mathematics education (Burkhardt, 1991), and, as noted earlier, this research is overwhelmingly supportive of the potential of calculators and computers in teaching mathematics. Moreover, other current research in mathematics education, though not necessarily directly concerned with the uses of technology, nevertheless also gives support to the value of technology in mathematics education. For example, there is considerable current interest in constructivist approaches to mathematics education. This is particularly relevant to a pedagogy that emphasizes the use of computers in the classroom because such a pedagogy would emphasize the notion of "discovery" learning, which fits in very well with the constructivist ethos (Davis, 1990).

A crucial question about the interrelation of research and practice is this: Can major curriculum reform be contemplated before there is a much larger body of research than at present that appears to validate a significant revision of the curriculum and a major change in the way mathematics is taught in the classroom? Here are several possible answers:

1. Yes, because the current situation is so bad, at least in the United States, that no changes of the kinds contemplated here could possibly do significant harm to children. This answer is, indeed, too flippant, but at least it conveys the idea that when you are near the "pessimal" point in a system, any perturbation is worth considering.
2. Yes, because much of the research needed is impossible except in the context of trying new things in mathematics classrooms. True, this answer implies only that there should be much controlled experimentation and many pilot programs. Still, no one should underestimate the potential resistance of all the communities affected by any attempt to make the kinds of changes envisioned here. We need Benezets but unfortunately the fertile ground that he ploughed hardly exists in the United States today.
3. Yes, finally, because of an intuition so strong that this is the right way to go that plunging ahead is the only way we can truly be fair to this generation of children. A statement of faith, indeed, but no evidence exists even to suggest that this is the wrong way to go, and considerable evidence exists already that it is the right way to go.

Understanding Mathematics: Technology and "Knowing Is Doing"

In a recent piece criticizing various NRC (1990) and National Council of Teachers of Mathematics (NCTM) (1989) reports, particularly their support for the use of calculators and computers in teaching mathematics, Edward Effros (1990) says: "As in any other language, drill and practice remain the most important tools at our disposal for learning the first principles of mathematics." This is really only a restatement of the familiar principle that "knowing is doing," by most professional mathematicians and most mathematics educators (Ralston, 1988, 1990). It reflects a belief that large portions of mathematics, certainly much of elementary school, high school and lower division college mathematics, can be learned only by doing lots of mathematics. The question is: Doing what? Calculations and symbol manipulation? Or interpretation of graphs, induction proofs and searching for patterns in data? No doubt both, but it is important to get the mix right. Right now, the emphasis on calculations and symbol manipulations in all of K–12 mathematics, but particularly in K–6, is so great that there is little "doing" of any other kind.

In any case, it is right to ask: What does the knowing-is-doing principle imply about curriculum change and about the use of computers and calculators in mathematics education?

Effros believes that "many of our students have already become *calculator idiots*" (emphasis in original). If, as claimed in the NCTM *Standards* (1989) "the current goal in most elementary school classrooms is far in excess of what is needed for tomorrow's society," why, asks Effros, "is it that many of our students are unable to do even the simplest pencil and paper calculations?" Effros's concerns deserve serious attention because his opinions are those of a significant portion of the mathematics research community.

Let's accept that drill and practice are necessary in order to learn large portions of school mathematics, where *learn* implies the ability to deploy, to use in problem solving, and to be able to apply in the further study of mathematics. *Learn* need not imply significant skill in calculation or symbol manipulation except insofar as such skill is necessary for the application of mathematics or the pursuit of further mathematical knowledge. Saying this implies the way out of Effros's dilemma.

Surely the kinds of drill-and-practice exercises—long sequences of calculations—that are the staple of pencil-and-paper instruction have no place in a calculator- or computer-based curriculum. But it does not follow, for example, that calculators do not lend themselves to a drill-and-practice regimen. On the contrary, it is possible to envisage calculator exercises in which students perform considerable numbers of operations to search for patterns or to draw inferences about how the various arithmetic operations work. One of the results of such a regimen would be to change the balance between the *drill* and the *practice*—much less of the former and much more of the (problem-solving-based) latter. We need much more effort than has been expended thus far on the development of such exercises, and we need research into their effectiveness. But surely there is no a priori argument that such an approach cannot be at least as effective as traditional pencil-and-paper approaches.

One of the arguments sometimes made against the virtual abandonment of pencil-and-paper arithmetic and symbol manipulation instruction is that students will no longer be able to (if, indeed, they can now) give instantaneous answers to questions such as: "What is 9 × 7? or What are the factors of $x^2 - 9$?" But this argument is a non sequitur. Indeed, it is necessary that students be able to answer such questions instantaneously if they are to be successful in higher mathematical pursuits. But the abandonment of pencil-and-paper arithmetic does not imply there is no need for mental arithmetic and symbol manipulation skill. Indeed, in a calculator world such skill is actually more important than ever because estimation abilities are more important than ever when using calculators and computers (Is the result plausible?), and mental calculation is crucial to an ability to estimate.

Thus, mental calculation deserves a much larger role than at present in a calculator- and computer-based curriculum. Not only do the addition and multiplication tables need to be memorized, but students need to be able to perform mentally a variety of two-digit computations, including perhaps some two-digit multiplications. Here, too, knowing is doing, so I have no doubt that students will need extensive drill and practice, much more than they get now, in mental calculation. In secondary school this stricture applies also to a variety of symbol manipulation tasks, such as, for example, factoring simple quadratics and solving linear equations. Would it not be so that the effort required to learn good mental mathematical skills would pay dividends beyond the skills themselves through the necessary development of mental discipline?

(A new emphasis on mental calculation, it has been noted [Stephen B. Maurer, personal communication, 1990], would be to return to classical times when writing materials were scarce and expensive. One wonders if there was consternation among mathematics educators when paper and pencil became cheap and readily available. Would the use of this new "technology" mean that students no longer would really "understand" arithmetic? Just as no one argues this way now, so, not very many years from now, no one will make the corresponding argument about computers and calculators.)

It is a major challenge to mathematics educators to show that the "knowing is doing" ethos can be present in a curriculum that emphasizes the use of calculators and computers in and out of the classroom and deemphasizes myriad skill objectives, few if any of which are useful or important. Not only is there no evidence to suggest that this challenge cannot be met, but there is every reason to believe that the flexibility and versatility of modern calculators and computer hardware and software offer opportunities for creative drill and practice that will be far better for education and motivation than the drudgery that has always been the inevitable concomitant of pencil-and-paper drill and practice.

Technology and Mathematics Education: When Is Its Use Right, When Wrong?

The foregoing has been an enthusiastic paean to the glories and necessity of a technology-based mathematics curriculum and mathematics pedagogy.

But of course it does not follow that, ipso facto, using calculators and computers in mathematics education is a "good thing." Are there prior guidelines that can predict whether the use of technology in mathematics education will be effective or ineffective? I think there are, and I think at least some of these guidelines can be inferred accurately without the support of research, most of which remains to be done — and needs to be done.

1. Software for in-class use should be an integral part of the pedagogy, not just an add-on used to cater to the current glamor of computers. This does not mean that computers cannot sometimes be used to illustrate what was previously illustrated with chalkboard or overhead. But such illustrations should be carefully designed to be pertinent to what has preceded them. Relatively little software produced for out-of-class purposes will be appropriate in this context without some redesign or additions.

2. In-class software needs to be designed in concert with textbook writing. Software designed to be appropriate for in-class use but not keyed to a particular text will have to be very user-friendly as well as very general, in the sense of having many options, if it is to be effective for use with a variety of textbooks. Correspondingly, it will be very difficult to write texts that envisage a computer in the classroom without designing the software at the same time. Eventually we shall probably learn how to write textbooks and software independently that can nevertheless be used effectively in tandem. Economics will probably dictate this. But I suggest that for some time it will be desirable, if expensive, to have team-developed textbooks-plus-software.

3. Less needs to be said about the use of technology outside the classroom. There is already enough software expressly designed for this purpose (e.g., Schwartz & Yerushalmy, 1987) that the general principles of user friendliness and generally good design are fairly well understood. Still, it needs to be said that most such software has been designed to be used with today's curriculum. A zero-based curriculum would undoubtedly require additional software and also, perhaps, the redesign of even the best current software.

A good technological environment in and out of the classroom will enable teachers to use computers and calculators as their assistants in teaching mathematics as an attractive and dynamic discipline. Such an environment will foster the learning of mathematics as both an experimental and theoretical discipline.

Since the first computers became available thirty-five years ago, most applications to already existing systems (e.g., data processing in business) have first involved using computers to do what was previously done manually, without changing the essential design of the system. Only later, sometimes years later, has the system itself been redesigned to take full advantage of a technology that permits far more than just the conversion of manual to automatic systems. Thus, early business systems were just computerized manual systems, but finally the computer revolutionized the way business systems were designed. So, no doubt, it will be with education. But because the problems are so large and the need to solve them so important, we must try to get over the first stage rather more quickly than has often been the case in other spheres. Tour de force demonstrations that computers can differentiate and integrate are one thing; using the computer to help students understand what derivatives and integrals are is something quite different.

References

Benezet, L. P. (1935, 1936). The story of an experiment. *Teaching of Arithmetic, 24*, 8–9, 25; *25*, 1.

Burkhardt, H. (1991). *How can micros help in schools? The research evidence.* Shell Centre for Mathematical Education.

Clements, D.H., & Nastasi, B. K. (1992). Computers and early childhood education. In T. Kratochwill, S. Elliott, & M. Gettinger (Eds.), *Advances in school psychology: Preschool and early childhood treatment directions.* Hillsdale, NJ: Lawrence Erlbaum.

Davis, R. B. (1990). Discovery learning and constructivism. In R. B. Davis, C. A. Maher, & N. Noddings (Eds.), *Constructivist views on the teaching and learning of mathematics* (pp. 93–106). Reston, VA: National Council of Teachers of Mathematics.

Effros, E. (1990). Some thoughts on "Everybody Counts." *Notices of the AMS, 35*, 559–561.

Hembree, R., & Dessart, D. (1986). Effects of hand-held calculators in precollege mathematics education: A meta-analysis. *Journal for Research in Mathematics Education, 17*, 83–89.

Kuhn, T. (1970). *The structure of scientific revolutions*, 2nd ed. Chicago: University of Chicago Press.

National Council of Teachers of Mathematics. (1989). *Curriculum and evaluation standards for school mathematics.* Reston, VA: NCTM.

National Research Council. (1990). *Reshaping school mathematics.* Washington, DC: NCTM.

Ralston, A. (1988). Knowing is doing. *New York State Mathematics Teachers' Journal, 38*, 92–94.

Ralston, A. (1990). Calculators, Computers and mathematical education: A response to Edward Effros. *Notices of the AMS, 35*, 1252–1253.

Schwartz, J. L., & Yerushalmy, M. (1987). The Geometric Supposer: Using microcomputers to restore invention to the learning of mathematics. In I. Wirszup & R. Streit (Eds.), *Developments in school mathematics around the world.* Reston, VA: National Council of Teachers of Mathematics.

CHAPTER SEVEN

Learning to Use Children's Mathematics Thinking
A Case Study

ELIZABETH FENNEMA
MEGAN L. FRANKE

THOMAS P. CARPENTER
DEBORAH A. CAREY

Abstract

This chapter reports how one teacher learned to use research-based knowledge about children's thinking. It contains a brief review of the research-based knowledge about children's thinking that provides the structure for Cognitively Guided Instruction (CGI), as well as the results of prior CGI studies. These studies indicated that teachers who had knowledge about how children solve addition and subtraction problems changed their instruction so that the children learned more mathematics and became better problem solvers. The chapter reports on a four-year study of one teacher learning to use that knowledge. Vignettes of her instruction and interview protocols illustrate changes in instructional behavior of Ms. G.* The chapter describes: (1) her changed beliefs about the role of the teacher and mathematics in the curriculum, (2) the mathematics she taught, (3) the integration of mathematics with other subjects, and (4) how she gained and used knowledge about individual children's thinking.

*Ms. G. is Susan Gehn, a first-grade teacher in Cottage Grove, Wisconsin.

The research reported in this chapter was supported in part by the National Science Foundation under Grant Numbers MDR-8955346 and MDR-8550236, and by the National Center for Research in Mathematical Sciences Education, which is funded by the Office of Educational Research and Improvement of the U.S. Department of Education under Grant No. R117G 10002. The opinions expressed in this chapter are those of the authors and do not necessarily reflect the views of the National Science Foundation or of the OERI.

During the last decade, we have gained increased understanding of learners' cognitions in mathematics (Hiebert & Carpenter, 1992). We have robust data on the strategies many children use to solve addition and subtraction problems (Carpenter & Moser, 1982; Riley, Greeno, & Heller, 1983); we have insight into how children think about place value (Fuson, 1990); and we have identified many misconceptions children have about rational numbers (Behr, Harel, Post, & Lesh, 1992). Increasingly, it is being suggested that curriculum developers and teachers should make use of this knowledge (Case & Sandieson, 1988). We have chosen to address this issue by investigating whether teachers can learn how to use research-based knowledge of children's thinking in mathematics to assess their students' knowledge, and how to take this assessment into account when planning and implementing instruction. Our work suggests that teachers can do these things and that they can be important influences on instruction and learning. The behavior of teachers changes as their understanding of children's thinking increases, and this change in behavior results in better mathematics learning by their students (Carpenter, Fennema, Peterson, Chiang, & Loef, 1989; Peterson, Fennema, Carpenter, & Loef, 1989).

We have also found that facilitating teachers' use of children's thinking to make instructional decisions is not a simple endeavor. For many, understanding children's thinking is not easy, nor is the use of such knowledge immediately obvious. Before teachers can use children's thinking effectively as part of their own practice, they must first use it with children for a period of time, and they must believe that children's thinking is important (Peterson, Fennema, & Carpenter, 1988). Not all teachers use such knowledge with equal effectiveness, nor do all teachers use it in the same way. However, our research has convinced us that teaching and learning are improved when teachers acquire research-based knowledge about children's thinking that can be applied in their classrooms.

Research-Based Knowledge about Children's Thinking

Research about children's thinking in addition and subtraction is based on a detailed analysis of the problem space (Carpenter, 1985; Carpenter & Moser, 1983; Riley et al., 1983). In this analysis, addition and subtraction word problems are partitioned into several basic classes that include different types of action or relationships representing different interpretations of addition and subtraction. Distinctions are made among problems that involve joining or separating actions, part-part-whole, and comparison situations. Within each class three distinct problem types can be generated by systematically varying the unknown. This scheme provides a highly principled analysis of problem types such that knowledge of a few general rules is sufficient to generate the complete range of problems.

The problem-type analysis is consistent with the way that children think about those problems and solve them. When young children initially solve the word problems, they directly represent the action or relationships in the problems; those problems that cannot be directly represented cannot be solved. Children advance in their problem-solving strategies from this direct representation to various counting strategies, to the use of relationships, and finally to the use of memorized facts. This development of strategies is closely related to the problem types and to the size of the numbers in the problems. Research has clearly identified the major levels that children pass through in acquiring more advanced procedures for solving addition and subtraction problems, and models have been proposed of the procedural and conceptual knowledge underlying performance at each level (Briars & Larkin, 1984; Riley et al., 1983). Thus, the problem-type analysis combined with a knowledge of strategies that children often use to solve the problems provide a principled body of knowledge that is directly relevant to the mathematics that forms the basis for the mathematics curriculum of most first-grade classrooms.

Curriculum Development and Teacher Education

Our approach to helping teachers acquire knowledge about children's thinking in addition and subtraction is built on recent research on teachers' thinking (Clark & Peterson, 1986; Shulman, 1986a, 1986b). This research indicates that previous efforts at curriculum reform failed partly because reformers attempted to prescribe programs of instruction without taking into account the decision making of the teacher implementing the program (Fennema, Carpenter, & Peterson, 1989a, 1989b; Romberg & Carpenter, 1986; Shavelson & Stern, 1981). Teachers are thoughtful professionals whose knowledge and beliefs influence their decisions about what to teach, how to teach, and to whom to teach. If one wishes to have an impact on instructional behavior, it is necessary to give teachers access to relevant knowledge and to facilitate their decision-making processes about how to use that knowledge.

In the teacher education program that is the focus of this research, experienced teachers were given the opportunity to acquire information about problem types and solution strategies in addition and subtraction, to discuss the instructional implications with other teachers, and to design their instructional program in accord with their own beliefs and teaching styles (see Fennema & Carpenter, 1989, for a complete description of our inservice education program.) This program, as well as the resulting instructional programs of the teachers, we called Cognitively Guided Instruction (CGI).

Results of Prior CGI Studies

For about four years we studied teachers and CGI—as they learned it, as they began to incorporate it into their instruction, and as they became expert in its use. During the first year we conducted an experimental study with 40 first-grade teachers (Carpenter et al., 1989; Peterson et al., 1989). We assigned half the teachers to an experimental treatment (CGI) and the other half to a control treatment. We administered pre- and posttests to teachers and their students and observed the teachers as they taught mathematics over a four-month period. Briefly, what we found was that after participating in the four-week summer workshop, experimental teachers taught problem solving more and number facts significantly less than did control teachers who had not participated in the workshop. Experimental teachers used different instructional strategies, listened to their children more, and knew more about the problem-solving processes of individual students. Experimental teachers were more inclined to believe that instruction should build upon students' existing knowledge than were control teachers. Students in experimental classrooms exceeded students in control classes in number fact knowledge, problem-solving ability, reported understanding, and reported confidence in their problem-solving abilities. Despite the fact that experimental teachers spent only about half as much time explicitly teaching number fact skills as control teachers did, experimental students actually recalled number facts at a higher level than did control students.

The Case Study

This case study presents an in-depth description of one teacher who participated in the experimental study and with whom we have continued to work. We describe her classroom, her instruction as she began to incorporate knowledge of children's thinking into her decision making, and how her instruction changed as she grew in her ability to assess the children's thought processes and plan instruction based on them. We document the changes that occurred in her beliefs about mathematics, how it is learned, and the role of the teacher as she became aware of her children's thinking.

The Data Base

The data base from which we have drawn this description of a teacher learning to use children's knowledge in instruction is composed of information gathered over a four-year period. We have transcriptions of structured individual interviews (Years 1, 2, and 4), formal and informal observations of mathematics instruction (Years 2 and 3), and field notes from nonstructured individual interviews and group discussions (Year 3). From this set of data, we abstracted ongoing themes of change in her knowledge, beliefs, and behavior.[1]

The teacher who is the subject of this case study was one of six selected at the end of the experimental study (Carpenter et al., 1989). For all of Year 3 (September–May), we used case study methodology to document the ways in which the six teachers used knowledge of children's thought processes in mathematics instruction. During this year, a project assistant was assigned to each case study teacher and spent at least two hours a week observing mathematics instruction in her classroom and talking to her about how she used children's thinking. In addition, the senior researchers and the project assistants met with the group of six teachers on a monthly basis to discuss their teaching. During the discussions, we did not tell the teachers how they were to teach nor did we suggest that anything that they were doing was either right or wrong. The discussions were usually initiated by the investigators raising a question, such as, "How do you decide to use symbols when you are teaching problem solving?" Although we did not hesitate to tell them what we thought about teaching or mathematics if they asked, the group discussions usually took place among the teachers themselves as they talked about problems they had encountered or successes they had experienced. Following the case study year, we again interviewed Ms. G.

In the case study year, we indicated to the teachers that we regarded them as professionals who were making instructional decisions and that our role was to help them develop an understanding of children's thinking on the basis of what we knew from research. We informed them that their role was to help us understand how the research knowledge could be useful in instruction. There is no doubt that our interactions with the teachers, as well as their interactions with each other, had an influence on the development of their beliefs and the change in their instructional behavior. We were not simply neutral observers of what was happening. However, our involvement did not detract from the value of understanding how one teacher learned to use children's thinking in mathematics instruction.

Ms. G.

Ms. G. was an experienced first-grade teacher. When she first joined the CGI project, she had been teaching a first-grade, self-contained classroom in a rural school for seven years. She typically had 20 to 25 children in her room. Many of the characteristics of her teaching were present before she learned about CGI and remained comfortably in place as she learned to be a CGI teacher. She obviously enjoyed teaching, worked hard at it, and was continually studying to improve. Her classroom was well organized, the children worked well independently, and the atmosphere in the room was warm. There was mutual respect between Ms. G. and the children, who seemed as fond of Ms. G. as she was of them.

Ms. G.'s routines, though not rigid, were well established and were used to help the class function efficiently. Mathematics did not always begin in the same manner or progress systematically within a similar time frame, but the children knew what was expected of them and they were comfortable doing what Ms. G. had taught them to do. Most children were on task most of the time. Seldom were discipline problems an issue, as Ms. G. anticipated them and acted to prevent them from happening. During the first two years of the study, she taught mathematics daily for about thirty minutes immediately after the lunch break.

If one were to characterize Ms. G. before she became an expert CGI teacher, it would be as an active (Good, Grouws, & Ebmeier, 1983) or direct-instruction teacher whose class ran smoothly. She was in control of classroom activities and felt responsibility for what her children learned. During instruction she told the children exactly how they were to proceed with the various activities she had planned. Her explanations were thorough and clear. She reinforced correct answers and had children correct any mistakes they made. Her mathematics program was determined by her textbook, although she supplemented it with materials provided by the publisher and at the learning centers that she had created. The text provided only one kind of word problems, simple result-unknown problems, and there were very few of these. The following episode, which took place early in Year 3, is illustrative of Ms. G. as an active teacher. At the time she had participated in the CGI workshop but had had little or no interaction with us or with other teachers outside the workshop.[2]

> Ms. G. was working with her entire class and was leading a discussion preparatory to the children doing a worksheet by themselves. At the bottom of the worksheet were some exercises where children were to add and subtract one-digit numbers (number facts) given in vertical form. Ms. G. first told the children that they had not seen this before and might not know how to do the problems. (This was an idea that the children had not encountered in class before.) Ms. G. asked the children what " − " meant. One child responded "take away" and Ms. G. said, "Take away. Good! You can use your counters. If you can do it in your head, that's okay, but there is nothing wrong with using counters." Ms. G. then demonstrated the solution to 8 minus 3 on the board by writing the number sentence as it was given in the text, drawing 8 circles next to it, and then crossing three of them out. After doing this, with no additional explanation or questioning of the children,

she said, "Write your answers underneath the line." The children went to work on the worksheet and she circulated around the room in response to questions.

Ms. G. explained what was expected of the children by drawing the circles and crossing three out, she demonstrated how to solve the problem, and she was explicit about where the response to each item should go. However, there was no attempt to ascertain whether the children understood the task or to relate mathematics to real life, nor was there any indication that their thinking was important. The activity was determined by the textbook. The children's success or failure in doing the activity depended to a large extent on their understanding of Ms. G.'s explanation.

When we finished the case study, Ms. G. had learned to use children's thinking in her planning and to help direct her teaching. She continually assessed her children's understanding and used the knowledge of their thinking to help her make instructional decisions. Because her teaching became so complex, it is difficult to illustrate with one episode. Perhaps the best way to illustrate the change in her teaching is to quote some of her own words from the final interview in Year 4. She starts by describing what had happened in her classroom the previous week.

Ms. G.: I used to read a book to the kids, and never really thought that the kids could solve some of these problems. I guess I didn't realize what their abilities really were. But now when we read books and stuff we will actually solve [many problems related to what is in the book]. Like today we were doing George Washington. O.K. When he was born. How many years ago would that be? (I used to think that they couldn't figure it out so I would just tell them.) Now we actually sit and figure it out. Before I would have said, no, this is too difficult, they will never get it, and just pass by. But I think because I'm listening to the kids and I see what they are able to do, I am giving them the opportunity to do that (i.e., figure it out themselves). . . . And I try to incorporate it because I think it makes so much more sense to the children anyway. You know, as long as we are reading a book, let's solve some problems that are in that book and actually sort of blend it altogether.

Interviewer: You said that you listen to your kids more. Why do you do that?

Ms. G.: I think that the kids learn from other kids much more than they learn from me. I see them listening to the other children and they think, hey, that is a neat way. . . . When the kids explain it they explain it in kids' terms, and the kids understand it better than when they hear it directly from me.

Interviewer: How do you think that your teaching has changed, when you said that you listen to kids more?

Ms. G.: I am real selective in what I will do in the classroom because our district has lots of workbooks and I don't order any of those any more. I try to make everything more experience, where the children have the experience to work with something or to do something so that it stays in their minds and hopefully they will retain it more. I think I have done that in a lot more areas than just math. I listen more to what the kids are thinking now in all my areas.

Interviewer: Why do you think that those experiences that tie things together that kids relate to, why do you think that those make such a difference?

Ms. G.: Because the kids can relate to it. It's more meaningful to them. I mean if we are reading a story on George Washington and then we solve problems about George Washington they are going to remember that because they solved the problem. . . . They're never going to forget that because they worked all morning on solving how old he was. If I had just told them it wouldn't have been meaningful for them at all and so it is all making a clearer picture in their minds, I think, and if they are responsible for their learning, they remember it much better than if a teacher is telling them.

In the lesson described by Ms. G., the children were asked to do challenging mathematics (e.g., figure out how long ago George Washington lived), and Ms. G. gave them freedom to do mathematics they were capable of doing. The problem had meaning for the children, and Ms. G. had made instructional decisions based on what she knew and believed about teaching and children's thinking.

Becoming a CGI Teacher

When we first worked with Ms. G., the district's adopted textbook was the basis of her instructional decisions, and she explicitly directed her children's learning activities. By the end of the time we worked with her, she was basing her instructional decisions partially on her knowledge of children's thinking and understanding, and she was permitting the children to direct their own learning in part. We shall first examine some of the changes in Ms. G.'s instruction and then discuss what made it possible for her to change.

Instructional Behavior

In order to illustrate the change in Ms. G.'s instructional behavior, we report two examples of her teaching recorded about eighteen months apart. In both cases, Ms. G. asked children to solve problems. The first example was taken from field notes made in the spring of Year 2. At this time, she had taken the CGI four-week summer workshop but had had no further interaction with us. The second example is from a report on her teaching near the end of Year 3, the year in which we were studying the six teachers.

> *Ms. G.*: I made 8 pies for my husband and he ate 2 of them. How many were left? Everyone get out your counters and we will do it together. [Pause] Lay down 8 counters. [Pause] Picture them as pies and my husband eats 2 of them. How many pies do I have left? Write your answer beneath the line. . . .

> After making sure that each child had used his or her counters to model the eight pies, had removed two of the counters, had counted the remainder, and had written 6 in the appropriate spot, Ms. G. continued: "Nan has 5 candy hearts. Dan gave her 6 more. How many candy hearts does she have now?" She asked Jim to show the class how he solved the problem. Jim demonstrated his solution on the overhead projector. First he counted out 5 hearts, then he counted out 6 into another pile. Then he counted all the hearts to get 11. Ms. G. pointed out that Jim should have started with the larger pile and added on to get the answer. She said that kids should always start with the larger pile in adding on.

To solve the first problem, children were led explicitly through each step of the problem and told how to solve it. The role of the student was to listen and follow instructions. In the second problem, Ms. G. did not indicate that she thought John's solution strategy a good way to solve the problem or that his thinking was important. Instead, she provided the class with an efficient strategy—one suggested by her textbook. For both problems she directed the learning explicitly and implied that there was one best way to get the answer. Although she focused the lesson on solving problems, there was little evidence that she thought children's thinking or problem-solving strategies were important as she made instructional decisions. This episode reflected Ms. G.'s beliefs at the time about her role as a teacher, which she expressed during the structured interview at the end of Year 1.

> *Ms. G*: [I] sometimes do not do anything with what one child says because if I devote myself exclusively to that child, I have to ignore

the other children and they will not gain anything from what I am saying.

The following instructional episode took place during Year 3 of the study. Once again, Ms. G. was working with the entire class on problem solving.

Ms. G.: Alice collects stamps. She had 25 stamps and she wants to have 40 stamps. How many more does she need?

[This is a change-unknown problem, which is more complex than the simple result-unknown problem in the first episode. The numbers are also much larger.] Ms. G. wandered around the room looking at individual slates which each child had. One child had written down 46 rather than 40, so she corrected the number. She repeated, "She's got 25 and needs 40." She noticed that Morgan was puzzled, so she suggested that he use counters. She restated the problem for Tom and said: "How are we going to find the answer?" She then called on various children to explain how they arrived at a solution.

Tom: 20 + 20 is 40, so I just took 5 away.

Ms. G.: Super. Why did you take away 5? [Long wait time.]

Tom: Because he had 25 so that would be 15.

Ms. G.: Good! Judy?

Judy: I counted by nickels.

Ms. G.: I didn't think of that way. Wonderful!. Ben, would you show us how you solved it on the overhead.

Bob counted by ones from 25 to 40 and put a counter on the overhead for each number. He then counted the counters he had placed on the overhead.

Ms. G.: Did anyone solve it a different way?

Alice: I thought in my head; one 5 to get to 30; one 10 to get to 40; 5 and 10 is 15.

Ms. G.: Wonderful! This is remarkable! Susan?

Susan showed how she counted from 25 to 40 with her fingers. She counted fingers on one hand three times and said that was 15.

If the strategy that a child used was unclear to Ms. G., she would say something like, "I don't understand. Could you tell me another way?" She provided hints only when a child appeared to be confused. After about five problems had been solved, the children separated into groups to continue problem solving. Ms. G. worked directly with a group of several children who had appeared not to understand the problems. She gave them problems with slightly smaller numbers, and she had them think through the problems step by step. She did not tell them explicitly what to do but provided hints like "Can you use your fingers?"

In this episode, Ms. G. asked children to explain their thinking and listened to what they said. If she didn't understand, she continued to question them. The textbook was not evident; Ms. G. had decided what to do on the basis of what she knew about her children's thinking. She was not explicitly directing the children's thinking. Instead, she had selected problems that were appropriate to their abilities and was encouraging them both to report and to think about their own problem solutions. Instead of asking children to follow procedures that she told them to use, she was asking them to engage in problem-solving behavior.

These two instructional episodes are representative of Ms. G.'s teaching at two different points in time. The classroom looked much the same and teacher and children were comfortable at both times, but there were major differences in Ms. G.'s instructional behavior. The role of the teacher had changed. Ms. G. was no longer totally directing the children's learning. Instead, she listened carefully to children and then used their thinking in making instructional decisions. The children were learning mathematics differently as they solved more challenging problems.

The Role of the Teacher

Ms. G.'s belief about what teachers should do changed. During the Year 1 interview, Ms. G. had indicated to us that the teacher

> is the person who introduces the concept and a person that demonstrates the concept and makes sure the children understand how to do the concept and makes sure that the children have enough practice. . . . [Teachers] have to really explain. . . . They have to make sure that the children understand what they are doing. [The teacher needs to] make sure that [the pupils] are following through with their work and monitoring it to make sure that they aren't making mistakes. [Students should] do the best work they can do. If they need help, they should come up to ask me.

The first episode described here illustrates this belief. It demonstrates that Ms. G. was a conscientious teacher who believed that she was responsible for ensuring that all of her children learned, and that they learned if they listened carefully to clear explanations from the teacher. She was able to implement her beliefs effectively during instruction. Children practiced what she told them to and asked for help when they needed it. She gave explicit instructions and showed the children exactly how to do the mathematics she was teaching.

Throughout the time we worked with her, Ms. G. remained a conscientious teacher. She continued to feel strongly that it was her responsibility to establish an environment in which learning would occur. However, she no longer believed that learning would occur just because she told the children clearly what they were to learn. She began to be much more directive in her teaching, she listened more carefully to children, and she built instruction on what she perceived they were able to do. Her children's thinking became important to her. During the Year 4 interview, we asked her how her teaching had changed.

> *Ms. G.*: I listen to the kids more, for one thing. . . . I am so much more aware of what the kids can really do that before I would never have thought about—first grade trying multiplication and division . . . that just comes from [my knowing] they can do it. . . . Then when I gave the children the opportunity and saw how they really do . . . solve those problems, now I realize that they can do them and give them the opportunity to do it.

She also reported that she did less direct teaching. When asked if she told children how to solve problems, she replied:

> I don't really share strategies. I mean, they come up with more ideas than I could ever think of . . . if a child is struggling and really has no idea where to begin I might . . . give them a starting place (like, Why don't you start with three counters), but I don't give them the strategies. . . .

She came to believe that children don't learn by being told what to learn.

> I just don't think that is enough and I really think they learn . . . by being an active participant. [That] is better than having them sit there.

We asked her directly what she thought her role as a teacher should be and what her perception was of her class as a CGI classroom.

> Well, to offer the curriculum, you know, to give them the opportunity to explore these math problems. To sort of facilitate discussion, ummm,

to give them the manipulatives, to listen to them, to plan instruction according to what they are needing right now.

A CGI classroom [is] where you build on the math knowledge of your children according to what they know. . . . You know, you don't build on objectives that say they should be doing this, this, this, and this. You sort of take what they know and build on from there.

Ms. G. became much more aware that children learned from each other. She said:

[When the kids listen to each other they] understand it better than when they hear it directly from me. It makes more sense to them. . . . I really see a lot of learning going on by [children] listening to the other children, I really do. I mean I see some of the slower kids really picking up on concepts . . . really learning a lot from listening to the other kids.

We would be remiss if we did not report the change in Ms. G.'s enthusiasm for teaching mathematics. She reported that before CGI, she always taught mathematics right after lunch and that she "had to drink coffee to stay awake." Now: "I could go for math forever."

Mathematics and the Curriculum

When we began working with Ms. G., the scope and sequence of the mathematics program was determined by the textbook that had been adopted by her school district. The small district was committed to mastery learning, and teachers were expected to have their children master the objectives delineated in the textbook. During the first two years of the study, Ms. G. accepted this situation and felt responsible for seeing that the children learned what was covered in the textbook. Not only did the textbook prescribe the sequence of instructional topics, it also prescribed most of the instructional activities. During our interview with Ms. G. during Year 1, she reported that the textbook was used to guide instruction explicitly.

Ms. G.: I always teach them to start with the largest number and then to add on the smaller number. The reason I do that is because that's the way our math series presents it.

When Ms. G. was interviewed at the end of Year 2, the textbook was still an important influence on her instructional decisions, but she was beginning to supplement it and to rely on her knowledge of her children. When asked how she used the mathematics textbook, she responded:

> *Ms. G.*: I like our math textbook. I use it a lot, except they don't get into story problems until a little later in the year and they only usually work with two basic types, so that's where I have to improvise myself. . . . I do look at those to see if it's a good idea, but sometimes I just, I think from experience, I have learned in other ways that the children have picked up on the concepts in a way that is better.

By January of Year 3, Ms. G. was using her textbook only as a guide. She still felt responsibility for covering each objective in the book, but she moved on as soon as she felt her children had met the objective. As Year 3 progressed, Ms. G. felt less and less inclined to use her text. She would have the children do only one row on the page or a few specified problems. Completing a textbook page became a low priority. The time spent on story problems was extended, and often the textbook pages were completely skipped.

Near the end of Year 3, Ms. G. announced to us that she asked and had received permission from her principal to teach mathematics without a textbook the following year. Instead, he was going to give her the money that would have been spent on workbooks to use in any way she wished for her mathematics program. During Year 4, she told us, "I never even pick up my textbook." Instead the curriculum was determined by her knowledge of the variety of the problem types and her children's understanding. She planned the curriculum on the basis of the children's activities in other subjects and the children's abilities to solve problems. She decided that the textbook was too limiting and did not build on what children knew and could do. We asked her directly about her use of a textbook in Year 4, and she responded:

> I don't use my textbook anymore. I am responsible for the objectives that our district has set down, but I know those objectives and so I just go about teaching them in a different way, plus I get all those objectives down in about nine weeks of school and then I just go on.

Ms. G. was making instructional decisions, not according to what the textbook suggested was appropriate, but as a result of considering what the children in her classroom could do. The instructional method she used was based on her belief that children were capable of solving problems and would profit from sharing their thought processes with each other and with the teacher.

Mathematics Taught

The mathematics taught in Ms. G.'s classroom changed dramatically. In fact, during Year 4, it met the standards suggested by the *Curriculum and*

Evaluation Standards for School Mathematics (National Council of Teachers of Mathematics, 1989). One of the most obvious changes in her mathematics instruction was in the amount of time she spent teaching mathematics. This change began when Ms. G. added 15 to 20 minutes to her mathematics class. It progressed to the inclusion of math in other subjects and at other times during the day. By the end of the time during which we worked with her, Ms. G. would often devote more than an hour each day to math, and mathematics problems often appeared in lessons in other subjects.

From Procedures to Problem Solving

The emphasis in Ms. G.'s mathematics instruction changed. Children were asked to solve a rich variety of addition and subtraction problems as well as multiplication and division problems. Although Ms. G. still spent some time on arithmetic procedures, the majority of the time during mathematics class was spent by the children on solving problems.

During the first year, Ms. G. believed that addition and subtraction were basically characterized by drill on the number facts with occasional illustrations of the facts in simple problems, that children needed procedural rules to help them solve problems (key words), and that the ability of most first-grade children was limited to the numbers from 0 through 12. The following excerpt from the Year 1 field notes illustrates her beliefs at that time.

> At first grade I think they should be able to understand the concept of joining two groups of things together. . . . [Addition] is adding two numbers together to come up with an answer . . . I tend to stress, which our book tends to emphasize, little key words in story problems. . . . I really stress that the kids know their math facts . . . zero through twelve by the end of first grade.

Consider what she did with her Human Math Machine as she moved from an emphasis on drill to problem solving. The Human Math Machine was a piece of poster board arranged so that a child's face could be seen through a clear window. On the front of the poster board was a drawing of a computer, and there were slots for input and output. A child (who *was* the Machine) would stand behind the poster board and another child would put a problem into the input slot. The Machine would then put the correct response out through the output slot. During Years 1 and 2, the children would take turns putting a number fact into the Machine through the input slot, and the Machine would respond by returning a correct response through the output slot.

New rules were devised during Year 3. Each child made up two word problems, which had specified dimensions; for example, one addend must be 1, 2, or 3 (to encourage counting on); it had to have a joining action; and the second could be any number the child wished.

Problem solving rather than drill became the focus of this activity. The children were asked to formulate their own problems and to discuss the solutions with each other. At this stage, Ms. G. decided that most of the children in the room were ready to solve problems by counting on. Using the addends she suggested (1, 2, or 3) would encourage counting on. Then, however, the children could use her directions in a way that was appropriate to their own thinking. Ms. G. was relinquishing the total direction of the children's learning while at the same time maintaining her own sense of the direction in which she believed they should be going. The following, from the Year 4 interview, shows how her beliefs changed.

> I don't do near as much drill and practice, I mean I would say I have cut down about 95 percent of what I was doing in drill and practice . . . I do basically all problem solving and try to incorporate that into every topic that we do, like even geometry, time and money . . . I integrate my math more into my other subject matters than I used to.

Extending What Children Already Knew

Mathematics ideas were taught as they related to each other and to the children's lives. Children were not told that they were going to learn something new in mathematics. Instead, new ideas were introduced as extensions of what had already been learned. This transition is illustrated in the change in the way Ms. G. approached instruction in time and money. Before her involvement in Cognitively Guided Instruction, Ms. G. had always taught isolated units on money and time. When the textbook unit on money or time was to be presented, Ms. G. told her children that they were starting a new topic and initiated the unit in such a way that the children knew something was changing. The text was followed for both topics, and children were taught to recognize various coins, to make simple change, and to tell time on a traditional clock to the hour and the half hour.

During the third year, Ms. G. approached the teaching of time and money as an extension of the problem-solving activities the children had engaged in from the beginning of the year. She continued asking the children to solve addition and subtraction problems as they had solved them before, but now she used money and time as the context for those problems. The mathematical ideas used in the money and time units were extensions of what the children already knew rather than something new and different. She found that many children could already recognize the coins and could tell time to the hour and half hour. Her time and money problems became more complex than they had been previously. The children simply perceived that they were solving addition and subtraction problems with time or money as the context. They believed that the money or time problems were mathematical problems that they could solve in any way they wished.

Ms. G.'s work with place value further illustrates how she came to integrate the mathematical ideas she was teaching. During Years 1 and 2 of the CGI project, she introduced place value when the textbook did. She presented the place value unit as something new the children were to learn, and she had them do the activities suggested in the book. These activities were isolated from other mathematical ideas and were usually pictorial representations of groups of tens and ones for which the children were to write the appropriate symbols. Early in Year 3, Ms. G. expressed her discomfort with what she had done with place value previously. After much thinking and talking to the other case study teachers, she decided to postpone the formal introduction of place value as a unit. Ms. G. knew that most of the children in her class could count to 30 or 40; she knew that some had rudimentary ideas about place value, whereas others did not understand grouping by tens. She decided to let these latter children listen to explanations that involved place value and to struggle with the concept. She began working informally with grouping by tens as appropriate situations presented themselves when the children were solving addition and subtraction problems with larger numbers. During November of that year, we observed and recorded the following:

> Ms. G. read a separate result-unknown word problem with the numbers 25 − 12 = ? The children solved the problem in a variety of ways, including direct modeling and counting back. After Ms. G. listened to a number of responses, she asked the class, "If I were to show 25 on the overhead, how many groups of 10 would I need? How many ones do I need? How can I take away 12?" A child responded that they could take away a 10 and then they would have to take away 2 more because 10 and 2 more is 12. The process was demonstrated for all the children, and Ms. G. then asked them to do five more problems where the children could use place value in problem solving.

Thus, place value was taught in relation to ongoing problem solving. It was not ignored, nor was it taught incidentally. Ms. G. consciously planned problems and activities in which place value ideas could be explored, and she had the children focus on the ideas whenever it was appropriate. At the end of the year, Ms. G. reported that she had not taught a formal unit on place value, and yet she felt her children understood place value ideas better than any group of children she had taught previously.

Integration of Mathematics with Other Subjects

When an opportunity presented itself for the children to use mathematics in other subject areas or in classroom activities, Ms. G. helped them to see that what they were doing was the same mathematics they did in math class. Although not all of the problems came from the child's own real-life experiences (consider the George Washington problem), the problems were meaningful to the children. Following is an excerpt from the year 4 interview.

> *Ms. G.*: One day last week I spent . . . almost two hours where I had bought cookies for the classroom and they had to find out how many cookies were in each package because I told them how many were in a row, how many rows, then they had to decide how many cookies there were altogether and then how many cookies would each child get if I were to divide them up and it took us the whole afternoon to do that problem. . . . We were doing an activity that afternoon with snow sculpturing with our fifth-grade friends, and I had brought these cookies and I thought I knew I had to divide 'em up and I thought, hey, let the kids divide them up. That will be a good math lesson. . . . At the end of the day then we did divide the cookies up to see if they were right and to see how many were left over.

She also reported to us that she had selected the packages of cookies so that each package had a different number of rows of cookies and different numbers in each row.

Assessment

Assessment is one of the building blocks of Cognitively Guided Instruction. We believe that:

1. In order to build instruction on children's thinking, it is necessary for teachers to find out what children are thinking and what they already know.
2. It is necessary for teachers to have a structured framework to guide assessment that enables them to see how the knowledge that is gained can be used in planning instruction.
3. Assessment should be integrated throughout instruction.

The ways in which Ms. G. organized and conducted her assessments, as well as their purposes, changed as she learned to use children's thinking. During the first two years that we worked with Ms. G., she used paper-and-pencil tests provided by her textbook series to find out whether children had achieved the objectives listed in the text. These tests were given whenever she finished a unit in the book. She used the information gained through the tests, not to plan her teaching, but to report to parents and the administrator of her school which objectives the children had mastered.

At the beginning of Year 3, Ms. G. still relied on the textbook tests. Although she was beginning to understand what her children knew, she was reluctant to use her own assessment of a child's knowledge. Consider the following excerpt from our field notes in September of Year 3.

> Children were doing routine exercises which involved using counters to model simple join and separate problems with sums less than 10. The children appeared to understand the task perfectly and were performing routine procedures. Ms. G. kept the children on task by external rewards when they all correctly followed her instructions and modeled a problem. [During an interview with us after the class] Ms. G. was asked if she thought the children knew how to solve those problems before the class started. She stated that they did. There appeared to be no doubt in her mind that each child in the class could do the problems. When asked why she was having them do something which she knew that they could do so easily, she said that she wanted to cover the book to make sure that they did understand. She was emphatic that they understood the problems before she started instruction for the day but that she had to make sure they understood by covering the book page. She did not believe that she could make instructional decisions based on her own knowledge of her children.

Gradually, during Year 3, Ms. G.'s assessment of children's knowledge and how she used it changed. For a time she continued to do formal, paper-and-pencil assessment about every third week. However, this was supplemented with informal assessments, which were done as she asked children to report to the group on how they had solved problems. Eventually the formal assessments, except for the required unit tests, were totally eliminated, and Ms. G. came to rely on her interactions with the children. She came to use the time during which the class was solving word problems together to assess her students. She would walk around the room watching and listening to the students as they solved problems. As she walked around, she had in mind an idea of which children would have difficulty with a particular problem, and she made sure that she kept a close eye on those children. As she said near the end of Year 3:

> I observe them as I'm giving problems and walking around and see who is doing what and asking them what they are doing. When they come up to me I check their papers, I can see what they have done. I might say to them, "Show me how you did this." . . . I see kids looking at the clock or I see kids looking at the number line or I see kids with

fingers or kids using manipulatives or I see the strategies they are using. . . . I can just tell who is not understanding this.

She learned to ask questions to gain understanding of the processes they were using to solve problems, such as, "What were you thinking in your mind?" or, "How did you know when to stop counting?" Such questions and her acceptance of their answers not only encouraged children to report their thinking in greater detail but also provided important information to Ms. G.

Ms. G. learned to use the knowledge she gained from her assessment of children during classroom instruction. This was evident in the larger variety of problem types that she asked children to solve. Originally she used only the simplest problem types with sums less than 10 that were in her textbook. When she learned that children could solve the other types, she began to ask children to solve all problem types with a variety of numbers.

Adaptation of Instruction

An interesting result of Ms. G.'s increasing use of her knowledge about children gained during assessment was her planning and implementation of instruction for individual children. She had told us during Year 1 that since every child in her room had to gain mastery over the various objectives, she did not move to a new objective until every child had learned the previous material. The children who learned the slowest were determining the pace of instruction, and Ms. G. believed that every child was learning the same things. Ms. G. did not feel that adaptation of instruction for different children was necessary, and she said she did not plan any formal grouping. She almost always taught the class as a group.

As Ms. G. began to listen to children more carefully, she became increasingly concerned that many children could do more than she was allowing them to do. She resolved this by having them work in smaller groups where she could more easily find out what they were capable of doing and could plan instruction directly for them. The grouping was flexible, not static, and decisions were made about who was to be in which group based on what she perceived each child could do and what she had decided was important for each individual child's progress. She selected problems and instructional activities for each group that would challenge them and yet could be *done* by them.[3]

Assessment thus became an integral part of her instructional day. It was done continuously as she interacted with children. She came to understand what each child knew and what kinds of problems they were able to solve. Late in Year 3, she reported to us that these surprises (children indicating that they either could or could not do what Ms. G. had thought they could) were significant in her instructional planning. Although she was careful to check out whether the surprise accurately reflected a child's knowledge, she used this new knowledge about a child to help in planning for the next day.

It appeared to us that the information we had provided about problem types and solution strategies provided the framework that Ms. G. used in ascertaining what children knew. At the beginning of Year 3 she was still imitating the interview procedures that had been demonstrated in the original workshop. As her confidence in her own ability grew, however, she learned to ask a greater variety of questions that seemed uniquely adapted to each child. Her listening improved and she was able to phrase new questions based on what a child had just said.

She also expanded the domain in which she assessed children's knowledge to content beyond addition and subtraction. The discussion of place value reported earlier is a good example of this. Although her depth of knowledge of place value concepts was more limited, she was able to ascertain when a child understood grouping and whether or not that understanding could be applied to a problem solution.

A Cognitively Guided Instruction Teacher

By the fourth year of the study, Ms. G. had learned to use children's thinking as a basis for making instructional decisions. Her role as teacher had changed from the provider of knowledge to the one who facilitated learning by posing problems for the children to solve based on what she knew about their thinking. The children's understanding of what they were doing had assumed top priority for her, and although she still felt that procedures were important, she believed that procedural knowledge was acquired as children developed an understanding and ability to solve a large variety of problems.

At the beginning of the study Ms. G. planned and taught for the entire group. By the end of Year 4, Ms. G. had come to listen to each child individually rather than listen to them as a group, and she responded to them as individuals. She knew each child's preferences for solution strategies and used this knowledge to help guide her lessons. She worked at getting the children to listen to each other in the hope that they would learn from each other and pick up additional solution strategies.

The overall change in Ms. G.'s teaching strategies is illustrated by the following examples from our field notes. In these examples she was focusing on helping the children write symbols for addition and subtraction problems. During Year 2, we observed the following: Ms. G. was working with the entire class and wrote on the board: $3 + 4 = 7$. "What does this mean?" (pointing to the + sign)." One child responded, "Add," and Ms. G. asked the children to model 3 + 4 with their counters and to confirm that 7 was the correct answer. She then wrote another number sentence on the board and went through the same procedure. Ms. G. simply expected the children to say, "Add," and then she moved on to the rest of the number sentence. She gathered no evidence of their thinking.

Late in Year 3, we observed the following:

> The children were to do a worksheet which involved writing number sentences for some problems as well as solving them. Ms. G., in explaining the page to the class, said, "Let's review what we do in writing a number sentence." Ms. G. read a problem and Allison was called on to solve the problem. After she described her solution, Ms. G. wrote on the board 8 + 8 = 16 and asked, "What does this" (pointed at the +) "mean?" (The same question she had asked the year before). "What do you think it means?" One child responded, "Add on." Ms. G. started to give another answer, but instead she asked, "Can you think of another way to say what it means?" The children said "more" and "add." Ms. G. waited quite a while between each response. (The sequence took 12 minutes.) The children were really trying to come up with ways to tell Ms. G. what they thought the symbols meant. It didn't come easily, but the children were thinking and trying to use their own words to explain the + sign.

In the first episode Ms. G. merely asked for one word (*add*) and went on without attempting to see if the children understood what it meant. In the second episode, she was working with the same idea. She pushed to find out what the children understood. She let them struggle (12 minutes), and the children attempted to put into words a major idea that was difficult to verbalize. Ms. G. learned that they didn't understand very well and it was possible for her to make instructional decisions based on what she learned. Not only did Ms. G. learn from interactions such as this; the children also learned. They learned to believe that their thinking was important and they learned to reflect on their own learning.

Some Conclusions

Throughout the four years that we worked with Ms. G., we observed that she became increasingly able and willing to use children's thinking as an aid in making instructional decisions. This was not easy for her and it involved much hard work and thinking. She did not use the knowledge about addition and subtraction in exactly the way we had hypothesized that teachers would use it, but we are convinced, as is Ms. G., that the knowledge she gained via CGI made it possible for her to understand her children's thinking in a way that she had been unable to before. The knowledge allowed her to interpret what her children said and to modify her instruction accordingly.

Ms. G. learned the structure during the CGI workshop, but it was not particularly useful to her until she incorporated it daily into her classroom. Throughout the next four years, she grew in her ability to listen to children and to modify instruction on the basis of what she heard. The process was self-sustaining. As she learned to listen to her children, she became better able to hear them and to use their knowledge.

We believe that our work with Ms. G. and the other case study teachers, as well as our experimental studies, illustrates that teachers can consider individual children's thought processes as they plan and teach. We believe that the addition and subtraction research provided a framework that teachers can use to organize their thinking about children's thinking; it provided a rationale for selecting activities for the children and a context for interpreting what they did. The framework provided a structure that made information accessible and retrievable as it was needed. Without such a framework, considering children's thought processes in planning and teaching is much more difficult.

As we have worked with Ms. G., we have come to understand how one teacher has learned to use knowledge gained from cognitive science research. That knowledge has been valuable both to her and to the other teachers we have studied and their students. Other reports such as this will confirm, deny, or enrich our understanding of how research knowledge can be translated into practice and how it can be used to improve mathematics instruction for all children.

Notes

1. Year 1 is the year immediately preceding the experimental study. During this year we interviewed teachers and pretested them and their children. Year 2 is the year of the experimental study, when we formally observed teachers, interviewed them again, and posttested them and their children. During Year 3 we conducted the case study, and during Year 4 we formally interviewed Ms. G.

2. When text is indented, it has been taken directly from field notes. Words appearing in brackets have been added to clarify the meaning.

3. As this book goes to press, Ms. G. is teaching the entire group and modifying problems and expectations based on her knowledge of individuals.

References

Behr, M., Harel, G., Post, T., & Lesh, R. (1992). Rational number, ratio, and proportion. In D. Grouws (Ed.), *Handbook of research on mathematics teaching and learning* (pp. 286–333). New York: Macmillan.

Briars, D. J., & Larkin, J. H. (1984). An integrated model of skill in solving elementary word problems. *Cognition and Instruction, 1*(3), 245-296.

Carpenter, T. P. (1985). Learning to add and subtract: An exercise in problem solving. In E. A. Silver (Ed.), *Teaching and learning mathematical problem solving: Multiple research perspectives* (pp. 17-40). Hillsdale, NJ: Lawrence Erlbaum.

Carpenter, T. P., Fennema, E., Peterson, P. L., Chiang, C. P., & Loef, M. (1989). Using knowledge of children's mathematics thinking in classroom teaching: An experimental study. *American Educational Research Journal, 26*(4), 499-531.

Carpenter, T. P., & Moser, J. M. (1983). The acquisition of addition and subtraction concepts. In R. Lesh & M. Landau (Eds.), *The acquisition of mathematical concepts and processes* (pp. 7-44). New York: Academic Press.

Case, R., & Sandieson, R. (1988). A developmental approach to the identification of central conceptual structures in mathematics and science in the middle grades. In J. Hiebert & M. Behr (Eds.), *Number concepts and operations in the middle grades* (pp. 236-259). Reston, VA: National Council of Teachers of Mathematics.

Clark, C. M., & Peterson, P. L. (1986). Teacher's thought processes. In M. C. Wittrock (Ed.), *Handbook of research on teaching*, 3rd ed. (pp. 255-296). New York: Macmillan.

Fennema, E., & Carpenter, T. P. (1989). *Cognitively guided instruction readings*. Madison: Wisconsin Center for Education Research.

Fennema, E., Carpenter, T. P., & Peterson, P. L. (1989a). Learning mathematics with understanding. In J. Brophy (Ed.), *Advances in research on teaching* (Vol. 1 pp. 193-220). Greenwich, CT: JAI Press.

Fennema, E., Carpenter, T. P., & Peterson, P. L. (1989b). Teachers' decision making and cognitively guided instruction: A new paradigm for curriculum development. In N. F. Ellerton & M. A. (Ken) Clements (Eds.), *School mathematics: The challenge to change* (pp. 174-187). Geelong, Victoria, Australia: Deakin University Press.

Fuson, K. (1990). Conceptual structures for multiunit numbers: Implications for learning and teaching multidigit addition, subtraction, and place value. *Cognition and Instruction, 7,* 343-403.

Good, T., Grouws, D., & Ebmeier, N. (1973). *Active mathematics teaching*. New York: Longman.

Hiebert, J., & Carpenter, T. P. (1992). Learning and teaching with understanding. In D. A. Grouws (Ed.), *Handbook of research on mathematics teaching and learning* (pp. 65-97). New York: Macmillan.

National Council of Teachers of Mathematics. (1989). *Curriculum and evaluation standards for school mathematics*. Reston, VA: Author.

Oxford American dictionary. (1980). New York: Oxford University Press.

Peterson, P. L., Fennema, E., & Carpenter, T. P. (1988). Using knowledge of how students think about mathematics. *Educational Leadership, 46*(4), 42-46. Reprinted in January 1990 by Chronicle Guidance Publications, Moravia, NY.)

Peterson, P. L., Fennema, E., Carpenter, T. P., & Loef, M. (1989). Teachers' pedagogical content beliefs in mathematics. *Cognition and Instruction, 6*(1), 1-40.

Riley, M. S., Greeno, J. G., & Heller, J. I. (1983). Development of children's problem-solving ability in arithmetic. In H. Ginsburg (Ed.), *The development of mathematical thinking* (pp. 153-200). New York: Academic Press.

Romberg, T. A., & Carpenter, T. P. (1986). Research on teaching and learning mathematics: Two disciplines of scientific injury. In M. C. Wittrock (Ed.),

Handbook of research on teaching, 3rd ed. (pp. 850–873). New York: Macmillan.

Shavelson, R. J., & Stern, P. (1981). Research on teachers' pedagogical thoughts, judgments, decisions, and behavior. *Review of Educational Research, 51*, 455–498.

Shulman, L. S. (l986a). Paradigms and research programs in the study of teaching: A contemporary perspective. In W. C. Wittrock (Ed.), *Handbook of research on teaching*, 3rd ed. (pp. 3–36). New York: Macmillan.

Shulman, L. S. (1986b). Those who understand: Knowledge growth in teaching. *Educational Researcher, 15*(2), 4–14.

CHAPTER EIGHT

Mathematics as Procedural Instructions and Mathematics as Meaningful Activity
The Reality of Teaching for Understanding

PAUL COBB　　　　　　　TERRY WOOD
ERNA YACKEL　　　　　ELIZABETH McNEAL

Abstract

The primary purpose of this chapter is to consider what it means to teach for mathematical understanding. To this end, lessons that occurred in two different classrooms, one at second-grade and the other at third-grade level, are compared and contrasted. The lessons were selected because both deal with place value numeration and because both involve the use of similar manipulative materials. The focus of the analysis is on the explanations and justifications that occurred in the lessons because it is here that we can gain an insight into the teachers' and students' beliefs about mathematics. The relevance of this focus is first illustrated by discussing a small-group problem-solving episode. The two children had different understandings of what it means to explain and justify, and as a consequence they were unable to work together productively and learn from each other. The first of the two lessons

---

The research reported in this chapter was supported by the National Science Foundation under Grant No. MDR 885-0560 and by the Spencer Foundation. The opinions expressed do not necessarily reflect the views of these foundations. Several notions central to this chapter were elaborated in the course of discussions with Heinrich Bauersfeld, Götz Krummheuer, and Jörg Voigt at the University of Bielefeld, Germany.

analyzed illustrates what we call the school mathematics tradition. Here, the teacher unknowingly initiated and guided the development of mathematics as procedural instructions in her classroom. It was quite possible for students to get correct answers and thus be effective in this classroom by following instructions for manipulating symbols that did not necessarily refer to or mean anything. In general, the teacher wanted her students to learn specific instruction that she had in mind all along and implicitly delegitimized solutions that involved the construction of numerical relationships. The second classroom exemplifies the inquiry mathematics tradition. Here, in contrast to the first classroom, mathematical interpretations and solutions were eminently discussable. Students were effective when they could contribute to discussions in which they explained and justified their thinking. Mathematics as it was realized in this classroom with the teacher's guidance had an experientially real, manipulable quality. The teacher and students were in effect constructing a mathematical world of numbers and relationships, and it was this that gave meaning and substance to symbols. Rather than funnelling students to procedures or answers she had in mind, the teacher built on students' contributions by highlighting aspects of their interpretations and solutions that were significant for their future learning. It was by providing this subtle guidance while respecting her students' thinking in a risk-free setting that the teacher made it possible for her students to learn with understanding.

Most mathematics educators agree that the next few years offer an almost unique opportunity. In the wake of reports such as those of the National Council of Teachers of Mathematics (1989) and the National Research Council (1989), there is a growing consensus that mathematics instruction at all levels should be reformed to make it possible for all students to learn with understanding. The need for reform is, at times, almost tangible — one can feel it in the air. Our goal in this chapter is to clarify what teaching for mathematical understanding might mean in practice by comparing and contrasting two mathematics lessons that occurred in different elementary school classrooms. We will pay particular attention to the way that interpretations, solutions, and answers were explained and justified in the two classrooms, for it is here that we can gain some insight into the teachers' and students' beliefs about the nature of mathematical activity (cf. Wilkinson & Martino, Chapter 9, this volume). In particular, we will ask whether mathematics as it was realized in the classrooms was a matter of following rules or instructions for getting right answers, or whether it was an intrinsically meaningful activity. Before addressing this issue directly, however, we will consider an episode that occurred as two second-graders worked together to illustrate the relevance of focusing on explanations and justifications as children solve mathematical tasks.

Explanations and Justifications

As the sample episode begins, the two children, Anne and Jack, were working together to solve a task that involved finding the sum of five 12's. Anne had written five 12's in vertical, column format and had used the standard addition algorithm to arrive at her answer, 60. She explained her solution to Jack as follows:

> 2, 4, 6, 8, 10 [adds the five two's in the ones column]. You put the "oh" there (writes "0" in the ones place), the "one" there (writes "1" at the top of the tens column). 1, 2, 3, 4, 5, 6 (counts the six ones in the tens column). Do you agree that's sixty?

In Jack's view, "That's impossible."

Jack: Hold on. This is ten. Look. 1, 2, . . . 8, 9, 10 (adds the five 2's in the ones column). Ten.

Anne: One there (points to the "1" she has written at the top of the tens column). 1, 2, 3, 4, 5, 6. 60.

Jack: 610?

Anne: You don't listen.

Jack: I don't get this.

As the episode continued, Jack repeatedly challenged Anne's solution, thus requiring her to give further justifications. As he did so, he became increasingly specific about why he questioned the step in her method in which 10, the sum of the ones column, became an "oh" and a one.

Anne: 2, 4, 6, 8, 10. Now listen. Put that "oh" there.

Jack: Yeah.

Anne: The one up there.

Jack: That isn't "oh."

Anne: 1, 2, 3, 4, 5, 6.

Jack: That's not "oh." That's ten.

Anne: No.

Jack: That isn't "oh" though. That's ten. 1, 2, . . . 8, 9, 10.

Anne: Now listen.

In our view, Anne's responses to Jack were justifications rather than explanations because she was replying to his explicit challenges about her solution. More generally, a justification occurs when someone believes they understand our thinking and challenges it. In contrast, an explanation occurs when someone does not understand our thinking and, without calling it into question, implicitly or explicitly requests clarification. In the sample episode, Anne justified her solution by repeatedly describing the procedural steps of the standard algorithm. At least in the setting of arithmetical computation, it would seem that a procedural description of this type constituted a justification for her, and she concluded that Jack's continued disagreement indicated that he was not listening. On the other hand, from Jack's point of view, Anne had yet to justify why she wrote "0" and "1." As a consequence of their different understandings of what counts as a justification, the two children reached an impasse. Anne therefore sought the teacher's help and both children explained their interpretations of the conflict to her.

Anne: He doesn't—he doesn't get. He doesn't get that right.

Jack: [Simultaneously] I don't understand. All the two's add up to ten. She takes one away from the ten.

Clearly, Jack interpreted Anne's act of writing "1" at the top of the tens column as taking away one from the ten, and therefore considered that her solution was irrational. Anne responded to this challenge by again repeating her procedural description.

Anne: Two plus two. 2, 4, 6, 8, 10. I know that's ten.

Jack: It is ten.

Anne: You put the darn "oh" there. 1, 2, 3, 4, 5, 6.

Jack: [Simultaneously] Show me how you are doing it. I don't understand.

Jack's comment, "Show me how you are doing it," was an explicit request for an explanation. He now seemed to accept that Anne's solution was legitimate, possibly because the teacher had not challenged it. From Anne's point of view, however, she had justified her solution four times in response to Jack's challenges, and she therefore concluded that he was being uncooperative. Jack, in contrast, desired to learn with what is typically called understanding and did not consider that Anne's descriptions of how she manipulated numerals were adequate as either explanations or justifications. In another episode, he in fact characterized her computational methods as "mixing up a bunch of numbers." An explanation or justification was acceptable to him only if he could interpret it in terms of mental actions on numbers experienced as manipulable arithmetical objects.[1] These mental actions include those of combining numbers, decomposing numbers, and transforming numbers (e.g., transforming ten units of one into one unit of ten, which is itself composed of units of one). In this exchange with Anne, Jack was unable to give a numerical meaning to her procedural descriptions, and she was unable to do more than specify how she manipulated numerals. As a consequence, the two children were unable to resolve the conflict even with the teacher's assistance.

One of the primary thrusts of the current reform movement in mathematics education is the call for mathematical communication in general and for collaborative small-group work in particular. The primary rationale for these recommendations is derived from the well-documented research finding that students' differing mathematical interpretations and solutions give rise to opportunities to construct mathematical knowledge (e.g., Yackel, Cobb, Wood, Wheatley, & Merkel, 1990). The conclusion to be drawn from the sample episode is not that small-group work should be abandoned. Instead, Anne's and Jack's failure to communicate effectively and learn from each other indicates the kinds of difficulties that can arise when students have conflicting views about what counts as a legitimate explanation and justification. In addition, the discussion of this episode illustrates that by focusing on the children's explanations and justifications, we have gained some insight into their conflicting beliefs about the general nature of mathematical activity. As we turn to consider the two classroom lessons, it will become apparent that Anne and Jack are in fact representatives of two different classroom mathematics traditions, which, following Richards (1991) we call *school mathematics* and *inquiry mathematics*.

Procedural Instructions

The first lesson we will consider occurred near the beginning of the school year in a third-grade classroom, when the teacher and students were reviewing the place value interpretation of two-digit numerals and number words. It is one of 28 lessons that was video recorded in this classroom over a six-week period (McNeal, 1991). We should stress at the outset that our goal is not to criticize a well-intentioned teacher. The teacher was in fact a sincere professional who held the intellectual and social welfare of her students paramount. By all traditional criteria, including those derived from the effective schools research, she would be judged to be highly competent. Our goal is instead to understand a little better how students can come to believe that mathematics is a rule-following activity divorced from their out-of-school experiences as they interact with teachers who would honestly like them to learn with understanding.

As the lesson began, the teacher introduced the notion of grouping by completing an initial activity in which she circled groups of ten tally marks and established with her students how many tens there were, how many ones were left over, and how many tally marks there were in all. Later in the lesson, she passed out Cuisenaire rods of two sizes and established with her students that there were ten little blocks in each of the longer blocks. We can immediately note that these two activities and the teacher's use of manipulative materials indicates that she genuinely wanted her students to learn with understanding. Thus, our focus is not on her intent, which is above reproach, but on the characteristics of mathematical activity as it was realized in the classroom despite her best efforts.

The teacher introduced the first task that involved the use of Cuisenaire rods as follows:

> All right. I'm going to put this [a card] on the board. One of them is going to tell you how many tens I want you to find and the other one is going to tell you how many ones. I want you to *use* your ten rod and your one rod[2] to show me the number that I put on the, on the board over here. . . . All right? The first thing I want you to show me is three tens and five ones.

The teacher posed eight tasks of this type, each time asking the children to say what number they had made once they had put out an appropriate collection of rods. The children were able to give the desired responses when called upon, and this part of the lesson generally proceeded smoothly. The teacher introduced the third task by stating, "Now, be careful here, guys. I want you to show me seven tens and no ones." The children responded "Seventy," and she then requested an explanation.

> Now, boys and girls, look up here [writes "7" on the chalkboard]. Why didn't we just put seven? Because you have seven of these [holds up a long rod], you have nothing else. Why didn't we just put seven?

This question demonstrates that the teacher did not follow her text rigorously. She drew on her professional knowledge to base her teaching at least in part on her understanding of students' possible misinterpretations.

> *Stephen*: Because there's seventy um blocks.

Teacher: Because there's seventy blocks. That's right. This tells us [underlines the seven] that there were seven tens. It has to have a zero here [writes "0" after the "7" and underlines it twice] to tell us that there are no ones. O.K.? Very good.

Stephen's explanation did not make reference to the numeral but was instead based on a numerical relationship that he constructed between the two type of rods. He mentally transformed the seven longer rods (i.e., seven units of ten) into seventy smaller rods (i.e., seventy units of one). The teacher acknowledged that his explanation was valid but then gave an alternative explanation that stressed a rule for mapping from the digits of the numeral to the number of each type of rod. In doing so, her actions implicitly delegitimized Stephen's interpretation of the task as a relationship between numbers as arithmetical objects and indicated to students that the focus should instead be on the numerals.

This exchange between Stephen and the teacher was just one brief moment in the process by which the teacher unknowingly and despite her best intentions inducted her students into the view that place value numeration in particular and school mathematics in general is about following procedural instructions concerned with how and where to write numerals. We should stress that this conclusion is consistent with other activities in this lesson and, indeed, with the analysis of other lessons conducted in this classroom (McNeal, 1991). For example, in the last whole class activity the teacher and children worked through four textbook tasks, each of which involved a pictured collection of base ten longs and individual cubes. In each case, the teacher asked, "How many tens do you see?" "How many ones do you see?" and "What number is that?" The children answered these questions appropriately for the first two tasks. The picture for the third task showed three longs and six cubes.

Teacher: How many tens do you see? Monica?

Monica: . . .

Teacher: [Problem] number three.

Monica: Three.

Teacher: How many ones do you see? James.

James: Four.

Teacher: Not in number three, James.

> *James*: . . .
>
> *Teacher*: Six. And what number is that, James?
>
> *James*: Sixty.

Here there is clearly a conflict between James's and the teacher's understanding of the task. It is in situations of this type that we can gain compelling evidence about the nature of mathematics as it was realized in this classroom. The teacher could have asked James to explain his answer or could have challenged him to give a justification. However, the way they both acted in the remainder of the exchange indicated that, for them, doing mathematics was a matter of following procedural instructions.

> *Teacher*: Look at number three, James. How many tens do we see? [Moves toward him.]
>
> *James*: Three.
>
> *Teacher*: And how many ones?
>
> *James*: Six.
>
> *Teacher*: And what number is that?
>
> *James*: Sixty-three.
>
> *Teacher*: O.K., let's look. James. Look. James, look. How many tens do we see? Three tens. How many ones? Six ones.
>
> *James*: Thirty-six.
>
> *Teacher*: Thirty-six. Good.

What did James learn in the course of this exchange? Did he construct numerical relationships and thus conceptualize thirty-six as the result of combining units of ten and units of one? It seems more likely that he found a way to produce an answer acceptable to the teacher by focusing on the number words *three* and *six*. When he found that "sixty-three" was incorrect, he tried "thirty-six" as an alternative. More generally, his goal in this social situation appeared to be to find a way to produce the correct answer and thus be effective, rather than to construct numerical relationships between units of ten and units of one as arithmetical objects.

The two sample episodes we have considered, those involving Stephen and James, were typical of those that occurred routinely in this classroom day after day. In the course of these interactions, mathematics was realized as a collection of unalterable procedural instructions that had to be followed to produce correct answers. As further evidence for this conclusion, we note that there was only one occasion in the entire lesson, the episode involving Stephen, when a student was asked to give an explanation. In addition, the teacher routinely challenged students when they gave wrong answers but did not expect them to give justifications, and indeed none did so. This is reasonable in a classroom where the construction of relationships between numbers experienced as arithmetical objects is implicitly delegitimized. In the absence of numerical meanings, arithmetic is reduced to an activity that involves the construction of rules for transforming one string of numerals or number words into another. For example, the students could be effective in this classroom by treating expressions such as "seven tens and zero ones" as entities in themselves that can be transformed into other expressions such as "seventy" or "70." This was also the case when manipulatives were used; a numeral or number word was read as an instruction for making a collection of physical objects ("70" means "put out seven big rods and no small rods"). As a consequence of this absence of numerical reference, there was nothing to which the teacher and students could refer to explain and justify the procedural instructions. The students either knew the procedural instructions to follow or they did not, and that was all there was to doing mathematics. Some students might, like Stephen, have continued to try and make sense of the tasks by giving numerals and number words numerical meaning, but this was a covert activity conducted in the face of instruction that implicitly discouraged such thinking.

The general nature of mathematics as it was realized in this classroom is reminiscent of what Lave (1988) called "folk beliefs" about mathematics. Lave, Murtaugh, and de la Rocha (1984) illustrated the effects of these folk beliefs by first documenting how adults solved price comparison problems as they shopped for groceries in a supermarket. They found that the adults figured out the best buy more than 95% of the time by using their own meaningful arithmetical methods. Lave and her colleagues then presented similar problems in a school-like setting and found that only approximately 60% of the answers were correct and that the adults attempted to solve the tasks by remembering standard paper-and-pencil algorithms. When questioned further, the adults were actually defensive about their effective nonstandard methods and believed that using the algorithms taught in school was the objective and rational way to solve arithmetical tasks in any situation whatsoever.

In light of our analysis, it seems reasonable to suggest that the folk beliefs about mathematics implicitly guided instruction in the third-grade classroom as the teacher interacted with her students. The teacher's actions implicitly and unknowingly served to foster the development of these beliefs, which are deeply ingrained in our society in general and in the culture of elementary school mathematics in particular. This is not to criticize the teacher, who was merely doing her job to the best of her ability. Our point is, instead, that the folk beliefs about mathematics are held by most administrators, parents, and (it appears) textbook writers, as well as by elementary school teachers. Mathematics instruction is typically thought of as a process of inducting the next generation into this stance toward mathematical activity. Reform in mathematics education involves breaking this cycle, and it is for this reason that we will now consider the instructional practice of a teacher who attempted to do precisely this.

Mathematical Truths

The lesson we will analyze occurred midway though the school year in a second-grade classroom in which every lesson was video recorded for an entire school year. This particular lesson was selected because, like the third-grade lesson, it occurred in a whole-class setting and because it involved the use of similar manipulative materials.

The teacher introduced the first instructional activity by placing two base ten longs and eleven individual cubes on an overhead projector. She then showed this array to her students for approximately three seconds and asked them to describe what they saw. Most responded "31," but one child, Michael, when the teacher called on him, said that he had seen 30.

Students: Agree, disagree.

Teacher: All right, let's take a look. How are we going to figure this out?

We can note immediately that, in contrast to the third-grade students, these second-graders assumed without question that they could challenge each other's interpretations, solutions, and answers (cf. Davis & Maher, Chapter 2, this volume). The teacher did not merely tolerate this behavior but in fact had actively encouraged her students to make challenges of this sort.

Michael: I . . . 30. I think it's 30 [walks to the screen].

Student: How can he get 30?

Teacher: Let's listen to his explanation.

Michael: 'Cause 10, 20, 21, 22, 23, . . . , 31. I missed one.

Teacher: You're just off by one. Sure, that's O.K.

The task might seem straightforward, merely specifying the cardinality of a collection of base ten blocks. Nonetheless, it gave rise to a discussion in the course of which Michael assumed without question that he should justify his answer when challenged. The teacher indicated that his assumption was appropriate by saying, "Let's listen to his explanation." It would therefore seem that, in contrast to the third-grade classroom, mathematical activity in this classroom was eminently discussable.

We should emphasize that this conclusion is not based solely on this exchange with Michael or on the analysis of this one lesson. Even apparently mundane activities, such as a worksheet of two-digit addition and subtraction tasks posed in vertical, column format, gave rise to discussions in which students explained and justified their algorithms. In this classroom, the process of constructing increasingly efficient computational algorithms was itself a problem-solving activity.

A further difference between the two classrooms becomes apparent when we recall the exchange between the teacher and James in the third-grade classroom. James was ineffective in the classroom when he gave an incorrect answer and strove to find a way to respond appropriately. Michael, however, continued to be effective even though his answer proved to be wrong, because to be effective in the second-grade classroom meant to participate in discussions by explaining and, when necessary, justifying interpretations, solutions, and answers. The teacher's final comment, "Sure, that's O.K.," in fact explicitly legitimized his activity.

As the discussion continued, another child, Brenda, indicated that she had an insight she wanted to share.

> *Brenda*: [Goes to the screen]. One thing is . . . you know there is two ten bars that make twenty and then you count. You count these [indicates the individual cubes], you count these 1, 2, 3, . . . , 11. Since there's eleven [inaudible] you see then you one, two [points to the two ten bars], then take ten and that makes 30. And then you put that extra one for eleven and you get 31.

Although it had never been discussed explicitly, the students seemed to understand what it meant to have a mathematical insight. As Brenda's explanation indicates, an insight typically involved constructing numerical relationships that made it possible to solve a task in a more sophisticated or efficient manner. We can also note that the way Brenda spoke indicated that, for her, numbers were objects that had a manipulable quality (e.g., "you . . . *take* ten and that makes thirty. And then you *put* that extra one for eleven"). This way of talking was quite general and occurred throughout the school year even when manipulative materials were not being used. In general, mathematics was experienced as being real by the students as well as by the teacher. In this classroom, students did not follow an instructional procedure to transform one expression, such "three tens and a one" into another such as "thirty-one." Instead, Brenda described her mental acts of creating and combining units of ten and of one. More generally, as the school year progressed, the students increasingly assumed without question that a number such as thirty-one was composed of thirty and one, and that the thirty was composed of three tens. This was not something that they had to remember consciously, but instead was experienced as a truth about the nature of numbers—from their point of view, numbers were simply made that way. In this regard, the mathematicians Davis and Hersh (1981) observed that "mathematicians know that they are studying an objective reality. To an outsider, they seem to be engaged in an esoteric communication with themselves and a small clique of friends" (pp. 43–44).

The distinction between studying a mathematical reality and engaging in esoteric communication succinctly captures the contrast between mathematics as it was realized in the second- and third-grade classrooms and, more generally, between inquiry mathematics and school mathematics.

The teacher's response to Brenda's explanation further indicates that mathematics in this classroom was experienced as being real, as having a manipulable quality.

> *Teacher*: That's right. She is telling me that if I take all of these ones and squeeze them all together [aligns ten cubes] I'm going to make a ten bar or a ten, and so that's how she figured out that if you take all those ones and make a ten out of it then you have a ten bar. It's the same thing.

In redescribing Brenda's explanation in her own terms, the teacher legitimized Brenda's interpretation of the task. In addition, she emphasized the aspect of Brenda's solution that she considered especially significant with regard to her students' mathematical development, that of creating a unit of ten from units of one. It was in this way that place value numeration grew out of the students' meaningful mathematical activity with the teacher's initiation and guidance.

In the remaining 31 minutes of the lesson, the teacher posed four more tasks and her students gave a total of twenty explanations and three justifications. As was the case in the discussion of the first task, the teacher and students took it for granted that each task could be interpreted and solved in a variety of alternative ways, and these explanations and justifications provided the teacher with important information about the sophistication of the students' mathematical thinking (cf. Ginsburg, Jacobs & Lopez, Chapter 12, this volume). In the course of these exchanges, the act of creating units of ten itself became an explicit topic of conversation. For example, the teacher introduced the third task by briefly showing the children a collection of four ten bars and sixteen cubes. The cubes were arranged in collections of three, four, four, and five and were separated by the ten bars. She first asked Joan to explain her answer of 56. Joan replied, "Well, I know four ten bars make forty" and then continued as follows:

> *Joan*: Take all four of these [points to a collection of four cubes] and one of these [points to one cube of the second collection of four].
>
> *Teacher*: O.K. She said take off all of these and one of these [removes five cubes], and now what do you want me to do with them?
>
> *Joan*: Put them right here [points to collection of five cubes].
>
> *Teacher*: Put them right here [she places the five cubes as Joan directed]. O.K. All right, let's see what happens. O.K., what happens! Now what do we have, Joan? What did we make?
>
> *Joan*: That makes ten.
>
> *Teacher*: That makes ten. Now, what are we going to do?
>
> *Joan*: Well, I know I have five tens and we have six right there [points to the remaining six cubes], so we should have 56.
>
> *Teacher*: So it should be 56. Right. Do you agree with her?
>
> *Students*: Yeah. Sure.

The teacher's role was not limited to that of expressing Joan's mental acts on numbers in terms of physical acts on the cubes. Note, for example, how she attempted to indicate the significance of Joan's act of creating a unit of ten by saying, "All right, let's see what happens. O.K., what happens! Now what do we have, Joan? What did we make?" She was, in effect, providing a running commentary from the perspective of one who could judge which aspects of Joan's activity might be potentially productive with respect to the children's mathematical development.

Shortly after Joan had given her explanation, another child, Paul, asked the teacher to return the cubes to their original positions. Once she had done so, he began his explanation as follows:

Paul: I squished all these together over here [points to the two collections of four cubes].

Teacher: What two?

Paul: These two four's.

Teacher: These two four's. O.K., I like that word, I think I will have . . . Webster add that word to the dictionary. *Squish*. I like that word.

In this case, the teacher capitalized on Paul's contribution to the discussion to propose a way to name the act of creating a unit of ten. In doing so, she facilitated the process of mathematical communication and encouraged students to reflect on this act, thus further guiding their mathematical development.

School Mathematics and Inquiry Mathematics

Before summarizing the differences between the two classrooms, we would stress that there was one important similarity—both teachers were authorities in the eyes of their students. A crucial difference between the teachers concerns the ways in which they expressed their authority as they interacted with their students. In one case, the teacher acted as the sole validator of the children's responses as she initiated them into the folk beliefs of traditional school mathematics. It is the process by which these beliefs are reproduced in the course of school mathematics instruction that must be broken if we are serious about reform in mathematics education. In the other case, the teacher guided the development of a community of validators in her classroom and initiated her students into an inquiry mathematics tradition. In one classroom, mathematics was realized as the activity of following procedural instructions to get right answers whereas, in the other, it was realized as the activity of creating and acting on mathematical objects. In one classroom the students learned with understanding, and in the other they did not.

Notes

1. Jack was one of several children observed on a daily basis in the classroom for an entire school year. To be sure, his understanding of basic mathematical concepts such as number were his own mental constructions. However, his problem-solving activity consistently indicated that he *experienced* numbers as entities that had an existence independent of his thought processes. These arithmetical objects in the world of his experience were abstract in the sense that the number 17 is not the same as any particular collection of 17 items. On the other hand, numbers were manipulable for him precisely because he experienced them as things. Fischbein's (1987) analysis of intuition in mathematics includes a detailed discussion of this aspect of meaningful mathematical experience.

2. *Editor's note*: Many experts on Cuisenaire rods object to the language that the teacher uses here. They argue that each rod has a permanent *color* name ("red," "black," etc.) but must not have any number name, except temporarily. They would allow phrases such as "If we call the red rod 'one' . . ." but would never speak of "the one rod" or "the ten rod." This makes it possible to deal with halves ("If we call the red rod 'one' . . ."), with thirds (calling the light green rod 'one'), and so on. These experts would have preferred to see the teacher say: "I want you to use white rods and orange rods" instead of referring to "your ten rod" and "your one rod."

References

Davis, P. J., & Hersh, R. (1981). *The mathematical experience.* Boston: Houghton Mifflin.
Davis, R. B., & Maher, C. A. What are the issues? Chapter 2, this volume.
Fischbein, E. (1987). *Intuition in science and mathematics.* Dordrecht, Netherlands: Reidel.
Ginsburg, H. P., Lopez, L. S., & Jacobs, S. F. Assessing mathematical thinking and learning potential. Chapter 12, this volume.
Lave, J. (1988). *Cognition in practice: Mind, mathematics and culture in everyday life.* Cambridge: Cambridge University Press.
Lave, J., Murtaugh, M., & de la Rocha, O. (1984). The dialectic of arithmetic in the grocery shopping. In B. Rogoff & J. Lave (Eds.), *Everyday cognition: Its development in social context.* Cambridge, MA: Harvard University Press.
McNeal, B. (1991). *The social context of mathematical development.* Unpublished doctoral dissertation, Purdue University, West Lafayette, IN.
National Council of Teachers of Mathematics. (1989). *Curriculum and evaluation standards for school mathematics.* Reston, VA: Author.
National Research Council. (1989). *Everybody counts: A report to the nation on the future of mathematics education.* Washington, DC: National Academy Press.
Richards, J. (1991). Mathematical discussions. In E. von Glasersfeld (Ed.), *Radical constructivism in mathematics education* (pp. 13-52). Dordrecht, Netherlands: Kluwer.
Wilkinson, L. C., & Martino, A. Students' disagreements during small-group mathematical problem solving. Chapter 9, this volume.
Yackel, E., Cobb, P., Wood, T., Wheatley, G., & Merkel, G. (1990). The importance of social interaction in children's construction of mathematics knowledge. In T. Cooney (Ed.), *1990 yearbook of the National Council of Teachers of Mathematics* (pp. 12-21). Reston, VA: National Council of Teachers of Mathematics.

CHAPTER NINE

Students' Disagreements During Small-Group Mathematical Problem Solving

LOUISE CHERRY WILKINSON AMY MARTINO

Abstract

This chapter addresses children's disagreements during small-group mathematics learning. The significance of collaborative interactions among children, which includes their disagreements, has emerged as a prominent theme in constructivist approaches to children's learning. The research questions addressed by this study are the following: (1) How are disagreements initiated, maintained, and resolved by children at this age? (2) What is the relationship between children's mathematics achievement and their participation in disagreements? (3) What is the relationship between children's social roles and their participation in disagreements? One group of four second-graders (two male, two female) met three times, and their interaction about mathematics was recorded. Disagreements were identified, and the sequence surrounding the disagreements was transcribed and analyzed. For each disagreement, the following was noted: who initiated it and how, who participated in it and how, whether children demonstrated the validity of their positions, and whose answer prevailed when the disagreement was finally resolved. Analyses were conducted on the relationship of disagreements to gender, achievement, and social role, since the following information was available for each child: gender; standardized mathematics and reading achievement scores, peer nominations of social status; and teacher designation as above average, average, or below average in mathematics achievement. One significant finding that emerged concerns the qualities possessed by the children who exhibited high levels of mathematical skills and intuitions. Verbal students shared their mathematical insights with their classmates during group work. These students, who acted as mathematical "catalysts," freely initiated and participated in disagreements as well as group discussions. These were students who were generally well liked and were seen as competent students with leadership capabilities. Implications for both theory and practice are discussed.

The significance of collaborative interactions among students has recently emerged as a prominent theme in constructivist approaches to children's social and cognitive learning. Constructivism emphasizes the importance of students' communication. Analysis of students' communication with each other has the potential to provide some insight about their knowledge and how they are able to modify it in light of the questions, challenges, and arguments provided by other students. The specific analysis presented here focuses on children's communication and disagreements about mathematics problems during small-group learning.

A central tenet of the interactional orientation to learning and development is that children learn from a diversity of interactions with others. The core of inquiry, therefore, is the set of social processes through which cognitive and communicative abilities are acquired and have meaning.

Variations of this general view are found in specific theories. These include the work of the Russian psychologist Vygotsky (1962) (for interpretations of this work and its relevance to learning and education, see Wertsch, 1985; for sociolinguistic approaches and pragmatic accounts of language development and disorders, see Halliday, 1975, and Snyder & Silverstein, 1988).

Despite the divergent viewpoints of these theories, there is a shared notion that significant individuals in children's social worlds mediate their interaction with that world and thereby determine which aspects of experience they will be exposed to. An elaboration of this notion is that the ways in which people organize communicative events (i.e., how rules are structured for who can speak, when and where speaking can take place, and for what purposes) reflect a society's enculturation process (Heath, 1983, 1989). The origins and continuous development of human behavior are viewed as the product of interpersonal processes among adults and children. The direction of learning and development is then considered as proceeding from the interpersonal (social) domain to the intrapersonal (cognitive) domain as defined by the progressive internalization of mental processes. In the case of school-age children, students are regarded as active collaborators with significant others involved in their learning; only later do they become capable of being independent learners.

The roots of constructivism can be found in the differing views of Piaget and Vygotsky, both of whom saw a central social dimension to the genesis of all conceptual development. Piaget believed that peers' communication with each other is useful in supporting children's efforts to decenter their thinking, as they consider multiple perspectives of a problem. Furthermore, he regarded the specific processes of questioning, explaining, and reflecting that children engage in when trying to solve a problem, to promote higher intellectual functioning.

Vygotsky is the best known developmental theorist to highlight the centrality of social interaction in individual cognitive development. According to Vygotsky, cognitive development is the result of the gradual internalization and personalization of processes that were originally experienced in social interaction.

The Problem

Communication, particularly verbal, significantly influences the ways by which students form ideas. Depending on who talks, what types of inquiry are engaged in or are acceptable, and the range of acceptable topics, students have different experiences to draw on in their own construction of knowledge. Communication is an integral part of learning and allows students to develop a distinct voice and internalize critical thought that has been discussed by other students (Pea in Schoenfeld, 1987). When children talk about mathematical ideas, they can call on a variety of viewpoints and can receive feedback about the validity of their own ideas (Steffe, 1990). Communication that results in students talking to each other about how they solved their problems is most important for mathematical learning (Cobb, Wood, & Yackel, 1990; Carpenter, Fennema, & Peterson, 1987; Maher, Landis, & Alston, 1987). Researchers have found evidence that reflection on "the paths of solution" helps students refine and consolidate their ideas.

One environment that provides opportunities for students to talk and work together on solving mathematical problems is the small group, where questioning, challenging of ideas, disagreements, and resolutions can be all part of the ongoing group work (e.g., Lindow, Wilkinson, & Peterson, 1984).

Empirical Background

Recent empirical research has been directly concerned with students' communication with each other during problem solving. Previous sociolinguistic studies have examined children's disagreements (e.g. Boggs, 1978; Brenneis & Lein, 1977; Eisenberg & Garvey, 1981; Geneishi & DiPaolo, 1982). In these studies, the language that children use to initiate, develop, and resolve the disagreements reflects their social, cognitive, and communicative knowledge about negotiating consensus in the face of opposition. Previous studies, however, have not examined systematically the relationships between these factors and children's learning of social or cognitive skills.

In one study, Lindow, Wilkinson, and Peterson (1984) examined the role of elementary school students' disagreements in small-group communication during mathematics. Communication that followed a verbal assertion of disagreement about a mathematics answer was studied. Disagreements were identified from thirty-two videotapes of small-group interaction during a two-week mathematics unit on time and money. The relationship of mathematics ability and achievement (as measured by standardized tests and teacher-initiated assessment) was examined, with four variables coded for each disagreement: who *initiated* it, who *participated* in it, whether children *demonstrated* the validity of their position, and whose *answer prevailed* among the group of students at the end of the disagreement. The results showed that high-ability mathematics students had significantly more prevailing answers and demonstrations. Participation, demonstration, and prevailing answers were all positively related to students' attributions of the mathematical competence of their fellow students. Prevailing answers were also positively related to students' achievement. Thus this study lends some support to the constructivist position: that high-achieving children actively communicate with other children about mathematical problem solving. Whether this relationship is correlational or causal (and in which way) remains to be discovered. One of the major findings emerging from previous studies (e.g., Webb, 1984) is that receiving no response or a terminal response to a request for help is negatively related to student achievement. Therefore, the ability to produce effective requests — that is, those that result in receiving the information/action requested — is a crucial communicative skill for children if they are to learn in small groups.

Previous research by Wilkinson and associates on students' interactions during mathematics and reading small groups (e.g., Wilkinson & Calculator, 1982; Wilkinson & Spinelli, 1983) focused on the characteristics of effective speakers, those who obtain appropriate responses to their requests. We identified the following dimensions of requests produced by especially skilled students: direct requests that were on task, perceived as sincere and addressed to one designated student. Findings from these earlier studies also lend some support to the constructivist position. In another study (Wilkinson & Spinelli, 1983), we found that students' mathematics achievement was positively correlated with their ability to get other students to respond to their questions in small groups. We identified the key role of "taskmaster" that emerged in these student-directed groups. Taskmasters were defined as students who catalyzed on-task communication by consistently pacing other students in the group, helping the group to manage time efficiently to keep on track, resolving disagreements, and getting the task done on time (Peterson, Wilkinson, Spinelli, & Swing, 1984).

Constructivist researchers in mathematics education also have acknowledged the critical importance of students' active learning and construction of mathematical knowledge as the key to successful teaching (e.g., Cobb, Wood, & Yackel, 1990; Carpenter, Fennema, & Peterson, 1987; Maher, Landis, & Alston, 1987). Although approaches differ, some commonalities emerge from these approaches, including the following: (1) knowledge is constructed by students, (2) cognitive structures activate the process of construction; (3) cognitive structures continue to develop that are stimulated by mathematical problem solving (Noddings, 1990).

A great strength of constructivism is that it encourages educators to think critically about the teaching-learning process (Noddings, 1990). Those who believe in the principles of constructivism no longer look for simple answers but, instead, develop a powerful set of criteria to judge the value of a student's contribution. A central element in constructivist approaches is the opportunity for learners to actively solve mathematical problems with others. The work of Cobb and his colleagues has shown that children as young as 7 years can work together in cooperative situations, such as small groups, to solve problems. Their work reveals the potential for maximizing discussion during explorations of problems, which in turn allows students to take other views into consideration, a critical element for development, according to constructivists such as Piaget. Furthermore, Yackel (1987) has shown that with greater experience in problem exploration, children become less dependent on teachers for support and confirmation and develop greater persistence at accomplishing goals. Fennema and Leder (1990) have pointed out that group activities allow students to become more autonomous in their thinking, another critical element in their construction of mathematical knowledge. Finally, Lesh and Zawojewski (1988) provide evidence that mixing ability levels within small groups provides opportunities for novices to view and perhaps incorporate more sophisticated problem-solving strategies into their own repertoire.

Research Questions

The research questions addressed by the present study are the following:

1. How are disagreements initiated, developed, and resolved by children aged 6 to 8 years?
2. What is the relationship between children's mathematics knowledge, their achievement, and their participation in disagreements?
3. What is the relationship between children's social roles and their participation in disagreements?

The focus of the analysis presented in this chapter is on one group of second-grade students who interacted during several mathematics problem-solving activities in the fall of the academic year.

Overview of the Complete Data Set

In the complete data set, 56 second-grade students in three second-grade classes were placed in groups of four by their teachers for mathematics. The groups included both boys and girls and were mixed in mathematics ability (as assessed by standardized tests). The one group selected for detailed analysis in this chapter was randomly selected by the authors. These groups met four times for this entire lesson cycle, which concerned addition/subtraction, and nonroutine counting problems. The complete interactions of the group during group problem solving were videotape recorded and transcribed, and disagreements are identified. For each disagreement, it was noted who *initiated* it and how; who *participated* in it and how, whether children *demonstrated* the validity of their positions, and whose *answer prevailed* when the disagreement was finally resolved. Two kinds of analyses were conducted: (1) the relationship of these variables about disagreements to gender, achievement, and social role, and (2) the mathematical principles involved in the student's problem-solving. The following information was available for each child: gender, standardized mathematics and reading achievement scores, peer nominations of competence/affiliation/leadership (sociometric task) scores; and teacher designation as above average, average, or below average in mathematics achievement.

Methods

Subjects

The data for this study were collected at an elementary school in a suburban New Jersey community. This community is a working- and middle-class community, and many of the children who attend this K–8 district are of Italian-American descent. The four students in the group studied were second-graders, standard English-speaking children: one 7-year-old boy, one 8-year-old boy, and two 7-year-old girls. All were members of the same class for the entire academic year and had attended the same school for first grade. The students were placed in a problem-solving group for the task. This type of activity was used by this second-grade teacher for mathematics, as was true for these students in first grade. This group of students experienced extensive practice in problem solving in both first and second grade.

Setting for the Study

The study was conducted at a school in a middle-class community in New Jersey, which is the site of a teacher development project in mathematics. The goal of the project, which is in its seventh year, is to assist teachers in their understanding of students' mathematical thinking, and to create classroom environments that encourage students to communicate mathematically with teachers and peers. Hence, the children participating in this study have, since their entry into school, worked in classrooms that supported their construction of mathematical ideas.

This K-8 elementary school system has approximately 650 students, who come from a community that is composed primarily of blue-collar and professional families. It is a stable school district employing mostly tenured teachers who have participated in a teacher development project in mathematics; this includes the first- and second-grade teachers who participated in this study.

The underlying premise of the project is that learning is abstracted from problem-solving opportunities, which allow students to explore patterns and relationships in mathematics, make conjectures, and discuss their ideas and revise their thinking. In this elementary school, children work in classrooms that encourage mathematical conversations, open sharing of problem-solving strategies, and time for reflection.

Although the project does not impose a method of teaching, it does encourage teachers to create certain kinds of learning environments. Such environments need to provide children with the opportunity to be actively engaged with the material to be learned, and provide physical embodiments for children to build multiple representations of the meaning of concepts and procedures to be learned. Children are expected to analyze problems, interpret situations, plan strategies for solutions, and monitor their own problem solving.

This environment provides a challenging role for the classroom teacher, who is responsible for designing activities and posing questions that stimulate interest in mathematical exploration and also for assessing students' mathematical thinking while furthering the development of mathematics as a whole. The teacher provides examples, explanations, and counterexamples to students' individual inquiries (Maher, 1988).

This setting thus encourages two new aspects of social and mathematical development: (1) Individual students' insights can contribute to other students' solutions, and (2) as students share a variety of strategies to solve the same problem, individual students are able to see that many times there is more than one way to solve a problem (Maher, 1988). In these classrooms, students become less concerned with the volume of problems that they produce and more alert to the quality of their solutions (Yackel, 1987).

Task and Grouping

The group studied was one of five in this class, which was formed by the teacher for mathematics lessons. All groups were heterogeneous in the mathematical abilities of group members (as assessed by standardized math achievement tests, to be discussed). Groups worked independently of each other in the classroom during mathematics, which typically occurred at about 9:00 A.M. each morning and was about 45 minutes in duration. The teacher provided procedural instructions, distributed the necessary materials, and then allowed the group to work on the tasks for the day. The rules for this task were to "work together" in groups to achieve solutions to the set of non-routine mathematical explorations. The teacher mandated that each group member be responsible for checking other group members' work with their own, and that each group achieve a solution by group consensus. The teacher was available to groups as she was needed, and she intervened occasionally.

In this classroom, the heterogeneous mathematics groups remained stable during the course of the year with minor variations; it is interesting to note that the same students were placed in homogeneous reading groups that remained stable throughout the entire year. There seemed to be very few restrictions on language use within the groups. The teacher asked that students not become too noisy so that they would not disturb other students.

We interviewed the teacher regarding her decisions about grouping for mathematics. In general, she mixed ability levels, with each group having at least one high-, one medium-, and one lower-ability student. She noted that during the year, the groups remained essentially stable in membership, with minor adjustments. The following excerpt from her interview reveals her view about leadership:

> *Teacher*: No, I didn't [appoint a leader]. I think someone always naturally assumes the leadership of the group.
>
> *Interviewer*: Do the children select these people — these youngsters out as their leaders?
>
> *Teacher*: I think a child just takes over most of the time.
>
> *Interviewer*: And the other children permit it?
>
> *Teacher*: Right.
>
> *Interviewer*: But you don't do it?
>
> *Teacher*: No, I haven't selected leaders . . . I don't think I've ever done that. No.

It is interesting that leaders do emerge and act in certain ways, even though the teacher did not distinguish an official leader.

This teacher saw her role as a facilitator for group problem-solving activities. "I just sort of set them up [the groups], explain more or less what they have to do. And then I just walk around to the different groups and just stand there and watch what they're doing and maybe ask some questions. If I see someone having difficulty I try and get them on the right track."

This teacher of approximately twenty years' experience shared her reflections about using group work and manipulatives to teach mathematics:

Teacher: I enjoy teaching math much better because of it.

Interviewer: Yes.

Teacher: I feel I'm doing something right.

Interviewer: How nice.

Teacher: Before I used to think "there's a better way" . . . when I was a child I wish I had been taught this way. I would have had a better understanding and not had a fear of math because I did, through school . . . these children feel that math is fun.

Materials

The set of word problems for the lesson cycle were unfamiliar to this class of second-graders. These materials (see Appendix 9A) were selected by the research staff; eight were based on original designs of Karen Fuson (1988), and two were developed by Robert Davis and Carolyn Maher and their staff. To aid in the solution of these problems, the children had paper and pencil, stones, and Unifix Cubes. Consequently, the children were familiar with the problem-solving materials but not with this type of mathematics problem.

Several factors rendered the problems designed by Fuson as nonroutine and more challenging than the standard "take away" problem. When solving "compare problems," children must keep track of three quantities: a big number, a smaller number, and their difference in size. Thus, they must first identify the "big" quantity and the "small" quantity before they can compare them to solve the problem. To further distinguish this set of problems, there are three types of comparison subtraction word problems (Fuson, 1988): "more and less," "equalize," and "won't get." An example is: "David has 14 Lego blocks. Paul has 5 Lego blocks. How many more Lego blocks did David have than Paul?" According to Fuson (1988), these problems "mirror the complexities of addition and subtraction situations in real life." Therefore, if a teacher had been calling the symbol ($-$) "take away," it was important to tell the children that ($-$) also had other meanings. She also recommended that the teacher should not teach the children how to solve these more difficult problems by locating key words. In fact, these problems were selected so that the key word strategy will not work. Also according to Fuson, research has indicated that the use of surface strategies (such as key words or number size) interferes with good problem solving. With regard to creating an environment for solving these problems, the classroom climate should focus on mathematical thinking, comparing strategies, and expressing ideas.

Video Recording

The children were observed and videotaped on four mornings, and the taping of their mathematics lesson cycle was conducted each day in their regular second-grade classroom (approximately 45 minutes in duration). The tape was transcribed using standard English orthography and was viewed a minimum of 15 times so that notes about context could be added.

Sociometric Task and Interview

The students and their teacher were interviewed separately (each interview about 20 minutes in duration) after the lesson cycle was completed (see Appendix 9B). Students' perceptions of the most "competent" group member were assessed with a sociometric questionnaire. The sociometric interview included eight questions with multiple parts (Appendix 9B). In addition to the competence scale, several questions addressed leadership (e.g., "Who reminds the children in your group to check their answers?") and others pertained to affiliation (e.g., "Who is your best friend in your math group?"). The highest possible competence and leadership scores for each individual were 16 (i.e., four nominations from each of four group members, including the interviewed student).

Identifying and Coding Disagreements

Disagreement episodes are defined as the interaction that followed a verbal assertion of not agreeing with an answer or a step in arriving at an answer to a mathematics problem (Lindow, Wilkinson, & Peterson, 1984). The initiation of the disagreement was preceded by an *antecedent position*—an assertion about the answer or step in arriving at the answer to a math problem (e.g., "Number 4 is two thirty"). The antecedent position was either verbally expressed, as illustrated in the preceding utterance, or nonverbally expressed, as in writing on somebody's worksheet. The second component, or the *disagreement*, expressed a position that contrasted with the antecedent position and involved a verbal display of dissent. The verbal interaction, or *resolution moves*, that followed the disagreement move were related to the dissented issue and characterized by the participants' attempts to convince each other of the correctness of their respective positions. The disagreement ended when consensus was reached or when the interaction shifted to a new topic. Consensus was signaled by a verbal acceptance of the antecedent or contrast position (e.g., "I guess you're right") or by nonverbal acceptance (e.g., changing one's written answer). Similarly, lack of consensus was signaled either verbally (e.g., "You can keep your answer if you want to get it wrong"; beginning new topical talk) or nonverbally (e.g., disagreement move was ignored; students resumed working independently). Table 9-1 presents the coding system.

The second author reviewed the transcripts while viewing the videotapes in order to identify and extract all disagreements. The disagreement move was identified first and was defined as an utterance that expressed a position incompatible with the antecedent position through asserting an alternative answer, negating the antecedent position, or questioning the position expressed in the verbal or nonverbal antecedent event.

Disagreement episodes were coded in the following way:

1. *Initiator*: Individual who initiated the disagreement move.
2. *Participants*: Individuals who made at least one verbal move that was related to the dissented issue.
3. *Number of resolution moves*: All utterances during a speaker's turn at talking that addressed the dissented issue and occurred after the dissention move. A resolution move could overlap another individual's speech or receive no response. Examples include restating one's position, asking for clarification, and providing justification.
4. *Type of demonstration*: A specific type of resolution move that provided justification, evidence, or a reason for one's position; it went beyond simple restatement of a position. Demonstration moves were coded as (a) appeals to authority (e.g., "It's right because the teacher told me"); (b) pointing out evidence on the worksheet (e.g., "Dimes are these"); (c) reading part of the problem aloud ("The party started one-half hour ago"); counting aloud (e.g., "One, two, three, four, five"); or (d) asserting arithmetic procedures or processes (e.g., "But it's takeaway, the third one is").
5. *Prevailing answer*: An answer proposed by the initiator or other participant(s) that was eventually accepted as the correct answer by all the group members.
6. *Topic*: Disagreement episodes were coded by topic: (a) mathematics disagreement, or (b) other.

TABLE 9-1 • *Coding Disagreement Episodes*

Antecedent Event

Verbal antecedent: A verbally asserted answer or step in arriving at an answer to a math problem. This category includes adjacency pairs, such as question–answer sequences:

   a. "This one's fifty minutes."
   b. A: "Did you say five dollars and forty cents?"
      B: "Yeah."
      C: "It's thirty cents."

Nonverbal antecedent: A written answer or step in solving the answer exists on a student's page and is responded to by another in the disagreement move.

Disagreement

The disagreement move verbalizes a contrast position by asserting an alternative answer, negating the antecedent position, or questioning the position expressed in the verbal or nonverbal antecedent event. The disagreement move is coded into one of the following categories:

Negates or asserts wrongness of antecedent position: Includes simple negatives, negation of the antecedent position, and wrongness-assertions of either a person or position:

   a. "Unh-unh"
      "Nah"
   b. "It's not three dollars."
      "You can't have that thing to yourself."
   c. "You guys are all wrong."
      "That's wrong."

Asserts alternative position: Asserts a different position than that expressed in the antecedent event; includes contradictions preceded by denials:

   a. "Six is a dollar forty-four."
   b. "No, it's nine o'clock."

Questioning repeat or challenge: A repeat (or reading) of the antecedent position with rising intonation; or a challenge of the antecedent position indicated by intonational disbelief:

   a. "Two point six three cents?"
   b. "Whaddya mean 'O' point forty-four?"
   c. "Wait a minute . . ."
   d. "What??!"

Appeals to authority: A weak appeal involves the form "s'posed to" implying but not naming authority; a stronger appeal names the authority (i.e., the teacher):

   a. "You're not s'posed to do both rows."
   b. "She said to check after everyone finished four."

Resolution Moves

A resolution move consists of all utterances that address the dissented issue during a speaker's turn at talk. Coded from the speaker's, not the listener's persepctive, a resolution move may overlap another speaker's turn, and it need not be responded to. Incomplete utterances are included as resolution moves if a meaningful unit is expressed (e.g., "It's not four-thirty, it's—"). One- or two-word "fillers," including "O.K.," "Yeah," and "Let's see," are *not* considered as resolution moves unless, for instance, "Yeah" signals agreement with another's position. Resolution moves may be also nonverbal (e.g., head nods, erasure of answer, one person leaves group, etc.).

Demonstration

A demonstration is a resolution move that provides justification, evidence, or a reason for one's position; it goes beyond merely restating one's position or countering another's. There are four major categories of demonstration, listed in hierarchical order according to the degree of cognitive process implied. If a demonstration move contains elements of more than one category, it is coded into the "higher" category.

Appeal to authority: A weak appeal involves the form "s'posed to" or "have to," implying authority; a stronger appeal names the authority (i.e., the teacher):

 a. "S'posed to be two point three cents."
 b. "No, she even wrote it."
 c. "Yes we do, on th'other one, the last one we did."

Locates evidence: Reference is made to something on page with the content either specified or nonspecified:

 a. "Dimes are these [points]."
 b. "Lookit."
 c. "See, a hamburger and a large milk."

Reads/counts: Justification consists of reading part of the problem, citing a problem element necessary to solve it, or counting aloud:

 a. "You have a superburger, its' eighty-nine cents."
 b. "The party started one-half hour ago [read]."
 c. "One, two, three, four, five."

Asserts arithmetic procedure/process: An arithmetic procedure is cited as justification or a process is elaborated to explain how one arrived at an answer; includes pointing out a classmate's error:

 a. "But it's takeaway, the third one is."
 b. "That's plus."
 c. "Zero plus two is two, five plus two is three, two takeaway two is nothing."
 d. "You're using the hour hand."

Outcome

The outcome is determined by reaching consensus, by explicitly not reaching consensus (e.g., "Let's all keep our own answers"), by resuming working, or by shifting the topic of verbal interaction. The outcomes when consensus is reached are coded into one of four categories:

Consensus reached with antecedent position prevailing: The answer expressed in the antecedent event is accepted by the episode participants. Acceptance can be signaled verbally or nonverbally.

Consensus reached with contrast position prevailing: The position expressed in the dissension move is accepted; includes multiple expressions of dissent following the antecedent event.

Consensus reached through teacher intervention: Teacher intervenes or is called on to intervene and confirm the correct answer.

Consensual answer generated during episode: Neither the antecedent nor the contrast position is accepted; the accepted answer is arrived at during resolution.

Consensus not reached: Work continues with the dissented issue not being resolved; or, if resolved (e.g., "Let's all keep our own answers"), consensus is not reached.

Interrater agreement was computed for identification of episodes, coding of participation variables, and coding of the types of demonstration moves as the number of disagreements divided by the total number of possible disagreements. Agreement on identification of episodes exceeded 91%; agreements on the initiator, participants, and prevailing answer variables all exceeded 90%; and agreement on the categories of demonstration moves was 80%.

Four scores were computed for each student:

1. A *participation score* was obtained by dividing the number of episodes in which students participated (i.e., made at least one verbal move) by the number of episodes that occurred in their group.
2. An *initiation score* was obtained by dividing the number of episodes initiated by a student by the number of disagreement episodes in the entire lesson.
3. A *prevailing-answer score* was computed by dividing the number of times a student's position prevailed (when consensus was reached) by the total number of prevailing answers.
4. A *demonstration score* was the number of a student's demonstration moves divided by the total number of demonstrations for the group. The student's total number of demonstrations in all categories was used because of the small number of demonstrations that occurred with each category.

Results

Standardized Tests

The four subjects took standardized tests in first grade (April 1989, the Iowa Test of Basic Skills mathematics and reading components) and in second grade (April 1990, the Iowa Test of Basic Skills mathematics and reading components). From an informal ranking based on classroom mathematics performance as obtained from the teacher, all four were considered to be average in mathematics achievement; their scores on the Iowa Test in mathematics, which was administered to them the spring of the prior year, revealed that one boy, Micky, and one girl, Dava, were higher achievers in mathematics compared with the other two group members in grade one. By second grade, however, Samantha's achievement in mathematics exceeded that of both Micky and Dava. This fact suggests that Samantha was highly engaged in the problem-solving activities in class, which in turn was reflected by her mathematics achievement on standardized tests.

Sociometric Results

The following information was obtained from the sociometric interview conducted after the problem-solving activities. Steve noted that Samantha and Dava were the leaders and that the role of a leader was "to try to help" the other children when they "get stuck on a problem." He viewed the responsibilities of "nonleader" as "to make sure that they have the right answer." Therefore, although it was never formally stated, Steve had a definition for "leader" as well as a notion that Samantha and Dava were in charge. The rules for being a leader were "to help people who don't understand what a person is saying." Micky perceived Dava to be a helpful individual, but there was no leader, although Dava saw herself and Samantha as leaders because they were "intelligent, smart." Dava also elaborated on the role of children who were not leaders by stating, "If I'm absent . . . like one of them will take my place." Dava had very resolutely placed herself and Samantha in the leadership roles for this group.

It was very interesting to learn that although she rated herself and Steve as the students who were best at keeping the group on task, Samantha expressed that there was no leader in their group. In her own words, she replied, "We'd just go around the group." She explained that the rules for a leader are "to try to figure out a way to settle things." She was also asked about the nonleaders and replied, "Well, sometimes one would lead the problem, someone would figure it out, the other would have to see if it was right or wrong." For Samantha each person in the group has a job, and a leader is a mediator. Why did Samantha not see herself in the same light as the other children did? What might explain her perception of a democratic division of labor within the group? What makes this interview especially unsettling are the data that were collected on Samantha, both her powerful use of mathematics and her dominant role within this group of four. Is there a possible explanation of why Samantha replied as she did?

Samantha was asked who was the best math student in the group, and she stated that Dava was. She did respond that Dava did not get the highest grades in math. However, when asked who was the best person at getting the group back on task, she admitted that she was the best at this. Samantha saw the group as a democratic institution with no leaders, and indicated that members took turns to determine roles.

In the sociometric analysis, Samantha and Dava tied for the highest affiliation and competence scores. Both girls are well liked by their classmates, whereas Micky rated in the middle and Steve received the lowest score. Dava received a slightly higher rating for leadership capabilities than did Samantha (a difference of one point). This pair was followed closely by Micky and then by Steve. Thus, in all three categories both Samantha and Dava were rated the highest. (For more details, see Appendix 9B.)

Quantitative Results of Disagreement Episodes

Table 9-2 presents a summary of the disagreement episode analyses for four sessions of this group; Table 9-3 presents their scores on particular variables.

Several interesting aspects of the students' participation in disagreements can be noted. Samantha participated in 16 of 17 episodes, which gave her a participation score of .94. Dava participated in 14 of 17 episodes (.82); Micky in 11 of 17 episodes (.65); and Steve, the least active, in only 6 of 17 episodes (.35).

TABLE 9-2 • *The Number of Each Type of Behavior by Group Members During Math and Nonmath Disagreements*

		Group Member			
		Samantha	Dava	Micky	Steve
Math	Initiations	1	6	0	1
	Resolutions	42	19	18	3
	Demonstrations (total)	20	10	5	0
	Level 1	0	0	0	0
	Level 2	8	5	3	0
	Level 3	5	6	2	0
	Level 4	7	1	0	0
	Participation	7	8	5	2
	Answer prevails	3	1	0	1
	Explanations	3	1	1	0
	Procedural—actions	12	6	5	0
	Procedural—information	3	3	2	0
	Answer-checking	9	5	1	0
Nonmath	Initiations	6	1	2	0
	Resolutions	30	18	15	6
	Demonstrations (total)	8	6	4	4
	Level 1	0	0	0	0
	Level 2	7	5	3	1
	Level 3	1	1	1	3
	Level 4	0	0	0	0
	Participation	9	6	6	4
	Answer prevails	6	0	0	0
	Explanations	3	1	2	0
	Procedural—actions	8	8	4	0
	Procedural—information	1	4	4	1
	Answer checking	0	0	1	1

TABLE 9-3 • *Student Scores for Disagreements*

Group Member	Scores			
	Participation	Initiation	Demonstration Score	Prevailing Answer
Samantha	.94	.41	.48	.60
Dava	.82	.41	.31	.20
Micky	.65	.12	.15	0
Steve	.35	.06	.07	.20

Initiation scores indicate that Dava and Samantha each initiated 7 of 17 episodes, whereas Micky initiated two and Steve only one episode. However, a closer look at the data reveals that 6 of 7 episodes initiated by Samantha were not about mathematics but were related more to social negotiations with the group, including issues of fairness and coordination of tasks for the group. Of the 7 initiated by Dava, 6 were about mathematics. Thus, although Dava and Samantha initiated an equal number of episodes (and more than the two boys), they were episodes of a different nature. Moreover, although Micky initiated few disagreements, he was active in other aspects of the disagreement.

There were 151 resolution moves, with 82 about mathematics. In the mathematics episodes, Samantha made 42 of 82 resolution moves, Dava 19 of 82, Micky 18 of 82, and Steve 3 of 82. Thus, Samantha made slightly more than half the mathematical resolution moves. Likewise, in the non-mathematics episodes Samantha made 30 of 69 resolution moves, Dava 18 of 69, Micky 15 of 69, and Steve 6 of 69. Again, Samantha seems to be the dominant force in both the mathematics and non-mathematics episodes; often she entered the episode to settle any group disputes. Dava is the second most involved; Steve is not very involved.

Analysis of the resolution moves provides some further insights about the group dynamics and problem solving. All students except Steve used demonstrations significantly more for mathematics than for non-mathematics disagreements. Of the 59 resolution moves that were demonstrations, none were appeals to authority, 32 of the demonstrations exhibited a student referring to some object, 19 demonstrations involved a student reading part of the problem or counting, and 8 of the demonstrations asserted arithmetic procedures or processes. Of these demonstration moves, Samantha made about half the "reference to an object" demonstrations while Dava and Samantha each made about one-third of the "reading and/or counting" demonstrations. Also, 7 of the 8 "assertions of arithmetic procedures/

processes" were provided by Samantha, all involved demonstrations of mathematical insights. This analysis provides supportive evidence of the high quality of Samantha's demonstration moves (for further information, see Table 9-2).

Samantha's communicative competence is revealed in the following exchange with Dava, in which Samantha asserted herself and Dava claimed unfairness. Samantha responded that Dava was not acting "fair" because she always looked at Samantha's paper when they worked together.

Dava: Samantha, friends?

Samantha: I know you like me, but you've gone too far Dava. I let you off the hook too many times. I'm not the little girl in kindergarten who would go around saying oh sure you can do that, oh sure I don't care . . . well I do Dava . . . what if one of these days I have a paper and you think that I'm smarter than you and I'm not.

Dava: What's 5 × 5?

Samantha: 5, 5, 5, 5, 10, 15, 20, 25! [Taps her fingers on the table as she counts.]

Dava: What's 50 plus, times 50?

Micky: You gotta add it 50 times.

Samantha: 325 (after staring into space).

Dava: You see you are smarter than me! I didn't know that problem.

Samantha: How do you know that I have the right answer?

Samantha communicated clearly and eloquently her position that she no longer wanted Dava to depend on her for "the right answer."

Some data suggested that the group members did acknowledge Samantha's leadership. Of the 6 prevailing answers, 3 were asserted by Samantha, 2 by Dava, and 1 by Steve. Therefore, the group members seemed to accept her answers that included mathematical demonstrations.

The data show that Samantha was highly involved in virtually all aspects of the group dynamics, which was consistent with the fact that Samantha was the highest achiever in mathematics and had the highest scores on two of the three sociometric scales. Her explanations involved thought and mathematical insight, and she maintained a balance within the group at all times, with group members participating. Although she did not initiate many mathematical disagreements, she participated in all episodes and supported the group in achieving consensus.

Dava and Micky both participated frequently but gave fewer high-level mathematical demonstrations than Samantha. However, Dava initiated the majority of the mathematics disagreements. Steve was the least active member and demonstrated no higher order mathematical resolutions.

Examination of the student's use of *procedural requests* (requests for action/information about a procedure), revealed that a total of 61 procedural requests occurred, of these, 43 were requests for action, and 18 were requests for information. The requests for action were as follows: Samantha (20), Dava (14), Micky (9) and Steve (0). The 18 requests for information were as follows: Dava (7), Micky (6), Samantha (4), and Steve (1). Once again Samantha was the most involved, providing the greatest number of directives toward her classmates to keep them on task, while Dava requested the most information. All children except Mickey provided more procedural requests for action in mathematics disagreements than in non-mathematics, and all except Samantha provided more procedural requests for action in the non-mathematics episodes.

In the following example, we see that Samantha's use of directives as the group reached a conclusion of one problem instigated the group to begin a new problem. It is interesting to note the frequency with which Samantha used directives to keep up the pace of the group, as well as Dava's use of questions to obtain explanations and further information. The children were working on a solution to the following problem:

Mrs. Brunette ordered 8 pencils. She also ordered some erasers. She ordered one eraser for each pencil and she also ordered 5 extra pencils. What was her order?

Dava: Wait a second, don't you think that we should be adding the pencils with the erasers?

Samantha: No, because they're not talking anything about . . . listen to me . . . listen to this sentence . . . Mrs. Brunette ordered 8 pencils . . . it doesn't really have anything to do with the pencils. [reads the problem] She has 8 pencils, right? She has 8 erasers too, right? All of a sudden she has 5 extra . . . 5 extra pencils.

Dava: 5 extra pencils?

Samantha: Erase the whole thing . . . it's 5 extra pencils.

Dava: 9, 10, 11, 12, 13!

Micky: Write 13 and 8. I told you I was right! I told you I was right!

Samantha: Put 'em together, Dava.

Micky: Here.

Dava: You wrote it in.

Samantha: You get the next one [to Dava].

Dava: [to Micky] You weren't supposed to write that in.

Samantha: He doesn't get to do the next one. He has to lose a turn [hands out papers], take yours, take yours. In fact, it's Micky's turn to read.

To continue, how the students checked their answers was examined. Answer checking, defined as any re-exploration of a problem that involved more than a restatement of the answer, occurred 17 times. Samantha checked answers 9 times, Dava checked answers 6 times, and Micky and Steve each checked one answer. Again, Samantha frequently monitored other members in their efforts to reach solutions. The example of answer checking that follows related to a money problem that required the children to find a way to spend exactly 17 cents. Samantha explained her solution, but Dava was not convinced that she was correct.

Problem: Erasers cost 3 cents each, pencils cost 5 cents each, and notepads cost 11 cents each. Chris spent exactly 17 cents. What did Chris buy?

Micky: 17 cents.

Samantha: I'm going to use my stones for this one. If he buys . . . guys listen . . . and watch this . . . take out your stones, and take out 17 [everyone takes out their stones]. O.K., you should have three left in your box [Samantha knows that their teacher always gives them 20.] Look now, now say it's 5, 11 plus 5 take 11 of those plus 5 is . . . 12, 13, 14, 15, 16, 17. The answer is he buys a pencil and a notepad.

Dava: He couldn't buy a notepad because 11 and 5 equals 16.

The Cube Problem: An Example

Although the quantitative results of the student disagreements provided the reader with a measure of the degree of involvement and nature of discourse exhibited by various group members, it provided little insight into the quality of the mathematical thinking that the children were engaged in during these several sessions. To provide a richer understanding of the intricate link between group interaction and mathematical problem solving, we now turn our attention to one mathematical disagreement in great detail.

Chapter Nine • Students' Disagreements   155

This problem, about a cube, stimulated more interaction involving higher order mathematical reasoning than any of the other problems (eight separate "assertions of arithmetic procedures/processes" demonstrations). It was a base four cube problem, a counting problem that involved notions of surface area and volume, and the children were provided with base four blocks to obtain a solution (see Appendix 9A). The question was: *How many of the small blocks do we need to make a big block?*

Appendix 9C presents the transcript of this disagreement. The lesson opened with a debate between Samantha and Dava. Based on her verbalizations during the argument, Dava seemed to focus on a surface-area model of counting the cube's individual blocks. She made comments that support this conjecture, such as "Can we take the block and trace each of the sides?" and "Would you like to do it by tracing the sides?" She also began to build a wall of units around the outside of the cube, which indicates her focus on the exteriors of the cube. In contrast, Samantha wanted to trace one side of the cube, but with a different purpose in mind. She suggested, "Trace the box and then you'd be able to fill the box and figure out the problem." To settle this dispute, Dava and Samantha used an "odds and evens" strategy. Samantha won the shoot, so they began to do the problem her way. Dava, however, was not convinced that Samantha's strategy of filling in an outline of the base of the cube with units would work, and she indicated her displeasure by stating, "I don't want to trace it." Therefore, Samantha continued to explain the flaw in Dava's strategy. She replied, "You know, Dava, why you can't do it that way . . . because you have to work around in the middle . . . you can't pick . . . if you . . . if we are going to do it your way what would happen is you would be doing the outside and you wouldn't be doing the inside. What I'm saying is you have to get all the blocks . . . you were going around it . . . if you do that there is no way you can fit the blocks inside." On the basis of this explanation, Samantha seems to be discussing the differences between surface area and volume. The group continued to work on Samantha's strategy. At this point, Micky suggested that they use a 4 × 4 flat block to save time. Samantha rejected this by saying, "No, we have to use the little blocks." Samantha filled in the bottom layers of the cube with units and counted, "4, 8, 16 in a row." She began to work on the second layer. Her block-by-block strategy was quite painstaking, but she seemed to believe that she needed to use the small blocks to build the cube. Dava suggested a better way to fill the outline with longs instead of units, but this suggestion was rejected even by Dava. At this point the second author arrived and repeated that the children could use anything they wanted to solve the problem. Micky immediately grabbed four 4 × 4 blocks and began to count. Dava grabbed two 4 × 4 blocks and some longs, and Samantha continued to add units to her third layer. Micky and Dava quickly achieved 64 as the answer, and Samantha, although she never finished her own model, seemed to agree with Dava and Micky when she said, "It was 64?" She finished the activity by attaching the symbol 64 to her traced outline of the cube.

Throughout this episode, Samantha's use of language indicated that she understood that the cube was solid and that there were blocks in the middle that were not visible to the eye. Her use of the words *fill*, *middle*, and *inside* indicated her representation of the cube as solid as she compared her method to Dava's using the terms *around* and *outside* to describe Dava's model (utterances 13, 25, 27). To continue, Samantha clung to an inefficient and time-consuming strategy but realized the value of a better strategy when Micky solved the problem before she did (utterances 57, 59). Finally, Samantha monitored and showed concern for all the members of the group. When a dispute arose, she insisted that every view be considered, stating, "Let's decide a way we all want to do it," and views were put to a vote (utterances 9, 16, 18). On several occasions Samantha requested quiet and patience during counting times by saying, "Some of us take slower time than the other" and ". . . count quietly" to Micky.

Samantha's paper showed a traced outline of the base of the cube with the number 64 written inside the traced outline, as shown in Figure 9-1. She was the only member of the group to place the number 64 inside her box. Dava, Micky, and Steve all had the traced outline of the base of the cube with the number 64 at the top of their papers.

FIGURE 9-1 • *Samantha's diagram of the cube problem solution*

How many of the small blocks will we need to make one big block? Make a drawing to show how you found your answer.

Individual Analyses of Students' Work

Samantha

A careful examination of the videotape transcript in Appendix 9C reveals that Samantha, a dominant member of this group, seemed to have a mental picture of the cube as solid (having volume), which was what the question intended. She suggested that the group trace the outline of one face of the cube, fill it in with unit blocks, and count to determine the total number (utterance 13). Another indication that Samantha viewed the cube as a solid came when she explained to Dava that her (Dava's) method was incorrect because she "would be doing the outside" and she wouldn't "be doing the inside." When Dava suggested that the group count the boxes on the surface of the cube, Samantha explained that when one does it Dava's way, "you can't count the inside box" (utterances 25, 27, 29).

At this point, Samantha followed the directions literally, traced her outline of one face of the cube, and began to build the cube a block at a time. Eventually, when the group was reminded that they could use any pieces they wished to find the answer, she took a cube (which is one piece of wood) and lined it up alongside her constructed cube to monitor her building. Then she used units to construct the third level of her cube; finally, running out of units, she picked up several longs to complete her figure. Although someone else reached the solution first and her counting was off a bit, Samantha's actions and verbalizations suggested that she views the task as a determination of volume, and she tried very hard to explain this to other group members. In fact, even her diagram indicated this since she put the numeral 64 inside her traced box.

Micky

Micky was almost as vocal as Samantha. In the beginning he agreed with Dava's strategy for solving the problem, which was to trace the faces of the block to determine the answer (utterance 16). However, his problem-solving strategies did not support his initial choice of solution strategy. For example, once it has been settled that the problem would be done Samantha's way, he suggested a more efficient strategy of filling the traced box with one 4 × 4 flat (utterance 30). This was rebuffed by Samantha, who insisted that they use the single units. Then Micky asked Samantha, "You gotta make the whole box?" She affirmed this, and he continued to work (utterance 38).

Two interpretations of Micky's query are possible: (1) He was unclear about the procedure for making the box, or (2) he may have been asking if he had to make the whole box one unit at a time. The second alternative is more likely. The evidence to support this conjecture was Micky's frustrated tone of voice, as well as his gestures.

Micky continued to work on constructing the cube by building up layers a unit at a time until the group was reminded that they were allowed to use any of the materials to solve the problem. At this point, Micky quickly destroyed his original block-by-block structure and built the cube with four of the 4 × 4 flats while counting quickly to 64. Therefore, although his initial agreement with Dava's strategy indicated a different representation of the problem (the cube as hollow), the majority of his actions after this incident indicated that he too was considering the cube to be a solid figure (utterances 30 and 51).

Dava

Her first strategy was to ask the second author if she could trace the sides of the block and put a flat block on each face of the cube (utterance 3). She defended her strategy against Samantha's which was to trace one face of the cube and fill it in with the blocks. In the disagreement about which strategy was best, they voted and did odds and evens, with Samantha winning the dispute. Dava protested because she still did not want to tackle the problem Samantha's way (utterance 19). Samantha proceeded to tell Dava why her strategy wouldn't have worked, and although Dava responded in the affirmative it was not clear that she was listening to Samantha.

Dava grabbed the cube and said, "Wait, all you gotta do is count these" (referring to the blocks on the surface of the cube) (utterance 28). Samantha explained why this would not work (utterances 25, 27, 29), and Dava began to count the blocks on the faces of the cube. She counted the blocks on three of the faces, stopped, and put down the cube. She continued to build the model using Samantha's method. As she filled in the first layer, she stated, "O.K., we did the bottom" (utterance 34). This comment was somewhat difficult to interpret, because it could mean her representation of the cube was hollow, solid, or transitional at this point. Her next comment was, "Now all we gotta do is make the block upper now" (utterance 37). This could indicate realization that the cube was solid.

As the activity continued, Dava built the block as a solid using Samantha's method. Next she commented, "I made a better way to do it . . . use these" (she holds up a long of 4 units), but then she corrected herself, saying, "but you can't, we're not allowed." Thus, Dava attempted a more efficient solution but did not use it immediately because she thought that she was not allowed to. At this point, the second author intervened and told her group that they could use anything that they wished to solve the problem. Dava continued her construction, but now with more efficient pieces (4 × 4 flats on longs of 4 units). She arrived at her answer quickly saying "I did it! The answer is 64!" It was still not clear whether she understood that the figure was solid or whether she was just following Samantha's model.

Steve

Steve said very little and was not active. When the task was introduced, Steve picked up one of the big cubes. He listened as Dava asked whether the cube could be traced, and he began to trace the big block (utterance 3). The next instance of any activity from Steve came when he agreed with Samantha's strategy for solving the problem by commenting, "Yeah, wouldn't that be easier?" (utterance 14). Once Samantha's method was agreed on, Steve traced the block with the others and began to fill it with the units. At utterance 31, Steve counted the units in the bottom layer of his box, pushed them aside, erased something on his recording sheet, and replaced the cube on his paper. After a pause, Steve began to build again (near utterance 40). Twice he complained that Micky was taking all his blocks; then, around utterance 49, Steve placed the units back in the dish. The other children went on to complete the problem, while Steve was holding a cube, turning it over in his hands. Micky and Dava achieved the answer, and Steve wrote this down.

Again, on the basis of Steve's external cues there was very little to analyze, and it would be presumptuous (based on such limited information) even to make an educated guess as to his mental representation. In fact, the only statement that Steve made during the entire session that indicated that he understood the task was his comment that Samantha's method would be easier than Dava's (utterance 14).

Conclusion

One important issue raised by this study is: What qualities did the children who exhibited high levels of mathematical skills and intuitions possess? Students who were more verbal in the interviews shared more of their mathematical insights with their classmates during group work. Samantha made the following observations about her group during her interview: "They wouldn't say, no that's not the way to do it . . . they'd say O.K. we'll do it that way and next time we'll do it a different way."

The students who acted as mathematical "catalysts" freely initiated and participated in disagreements as well as group discussions. Other qualities that these students possessed (according to the sociometrics) were that they were generally well liked, with high affiliation scores (e.g., out of 9 possible points Dava and Samantha each received 7), and were seen as competent students with leadership capabilities (out of 16 possible points Dava and Samantha each received 13). Samantha and Dava both received high affiliation scores and were viewed as leaders by the other students. Both answers and solutions of these two students were generally listened to and accepted by the other group members.

This group with its "student-catalyst" had rich and long mathematical episodes—in other words, mathematical episodes that involved more than simply one student "stating" and the others "accepting" the solution. These student-catalysts freely question and initiate disagreements; they demonstrate and participate often (see Table 9-2). Samantha provided more resolution moves and generally high-level mathematical demonstrations (7 out of 8 high-level demonstrations) than the other students in her group. She also participated in 16 out of 17 dissension episodes and provided 28 out of 59 demonstrations during this series of problem explorations.

The following example illustrates how Samantha mobilized the group by disagreeing with Dava, but clearly opened a forum for the students to listen to different strategies and vote for the one that made the most sense. Notice these social moves when examining the following discussion of the cube activity.

> *Samantha*: Wait a second. Trace the box and then you'd be able to fill the box and figure out the problem.
>
> *Steve*: Yeah! Wouldn't that be easier?
>
> *Samantha*: Let's vote, who wants to do it my way? [Steve raises his hand.]
>
> *Samantha*: Who wants to do it Dava's way? [Micky and Dava raise their hands.]
>
> *Dava*: It's two against two. [Micky watches Samantha and Dava.]
>
> *Samantha*: Two against two . . . so we'll have to do odds and evens. Evens is mine, odds is hers. Ready Dava . . . one, two . . . [pause] Wait a second! You can't look at my number! . . . one, two, three . . . I won! We're doing it our way.

Although Samantha exhibited competence in problem solving, the data suggest that her role as resolver of mathematical conflicts within the group was important to her as well.

One of the intriguing findings from the study concerned the nature of the task. Clearly the cube problem produced a greater quantity of high-level mathematical communication (8 high-level demonstrations in the cube problem) and promoted more task-oriented discussion than the subtraction word problems (see Appendix 9A). We suggest several reasons that the cube activity could have been better suited to stimulate the level of communication that some believe is so valuable for children's development of knowledge.

First of all, it is possible that the word problems, with their heavy language component, may have obscured the real mathematics of the problem for children in this age category. Frequently, we observed that the children discussed one statement of the word problem and disregarded another piece of critical information. In the following example, one student seemed to be digesting the problem bit by bit, but not as a whole. The full problem is:

> Mrs. McGlade ordered 7 pencils. She also ordered some erasers. She lost 6 erasers. Now the number of erasers is the same as the number of pencils. What was her order?

The student replied: "Oh, it tells you the answer, the number of erasers is the same as the number of pencils." The student does not appear to understand that in order for pencils and erasers to be the same, 6 erasers must be lost. This piece of missing information changed the problem significantly for that student.

Another factor that might account for the success of the cube activity in stimulating communication is that it required the building of an appropriate model of a physical object, which gave the children a basis for comparison of strategies (Steffe, 1990). In contrast, the word problems didn't necessarily require visible model building at all, and if a model was built it was frequently of a very unnatural mathematical situation. The following problem illustrates this point:

> Mrs. Brunette ordered 8 pencils. She also ordered some erasers. She ordered one eraser for each pencil and she also ordered 5 extra pencils. What was her order?

The nature of this problem is strained in the sense that few people order 8 pencils and immediately afterwards order 5 extra pencils. Depending on the way a child processes this problem, it could be interpreted as an order of 8 pencils and 8 erasers, 13 pencils and 8 erasers, or 13 pencils and 13 erasers, which can lead to considerable confusion for a group of second-graders. Although the children did not have pencils and erasers placed at their desk to solve these problems, they were given access to a variety of manipulable materials to aid in the solution of these word problems.

A third explanation of why the cube problem might have promoted more mathematical discussion could be that it naturally required the children to use the materials given to them in order to provide a convincing explanation of their solution strategy. This model building provided time for reflection and refinement of one's strategy for solution (Steffe, 1990). An example of this occurs in the transcript of this problem. While Dava and Samantha argue over surface area and volume, Micky improves on Samantha's strategy of building up the cube one block at a time, by using four 4 × 4 flat blocks to reach the same goal. The nature of the task provided time for model building and discussion because the solution was not immediately obvious.

Finally, it should be noted that the base four cube activity produced: (1) a high number of demonstration moves to reach a group consensus,

(2) the majority of the highest level mathematical demonstrations, and (3) consistently high degrees of answer checking (see Table 9-2 for details).

References

Boggs, S. (1978). The development of verbal disputing on the part of Hawaiian children. *Language in Society*, 7, 325–344.

Brenneis, D., & Lein, L. (1977). "You fruithead": A sociolinguistic approach to children's dispute settlement. In S. Ervin-Tripp (Ed.), *Child discourse* (pp. 139–177). New York: Academic Press.

Carpenter, T., Fennema, E., & Peterson, P. (1987). Cognitively Guided Instruction: An alternative paradigm for research in teaching. *Proceedings of the ninth annual meeting of PME-NA*, pp. 225–230.

Cobb, P., Wood, T. & Yackel, E. (1990). Classrooms as learning environments for teachers and researchers. In R. B. Davis & C. A. Maher (Eds.), *Constructivist views on the teaching and learning of mathematics*, pp. 140–168.

Eisenberg, A., & Garvey, C. (1981). Children's use of verbal strategies in resolving conflicts. *Discourse Processes*, 4(2), 149–170.

Fennema, E., & Leder, G. C. (Eds.). (1990). *Mathematics and gender*. New York: Teachers College Press.

Fuson, K. (1988). *Addition and subtraction word problem booklet*. Unpublished manuscript, Northwestern University, Evanston, Illinois.

Geneshi, C., & DiPaolo, M. (1982). Learning argument in preschool. In L. Wilkinson (Ed.), *Communicating in the classroom* (pp. 49–68). New York: Academic Press.

Halliday, M. A. K. (1975). *Learning how to mean: Explorations in the development of language*. New York: Elsevier.

Heath, S. B. (1983). *Ways with words*. New York: Cambridge University Press.

Heath, S. B. (1989). The learner as cultural member. In M. L. Rice & R. L. Schiefelbusch (Eds.), *The teachability of language* (pp. 333–350). BBaltimore, MD: Paul H. Brookes.

Iowa Test of Basic Skills. (1989, 1990). Chicago: Riverside Publishing Company.

Lesh, R., & Zawojewski, J. (1988). Problem solving. In T. Post (Ed.), *Teaching mathematics in grades K-8: Research-based methods*. Boston: Allyn and Bacon.

Lindow, J., Wilkinson, L., & Peterson, P. (1984). Antecedents and consequences of verbal disagreements during small-group learning. *Journal of Educational Psychology*, 78(5), 334–340.

Maher, C. A. (1988). The teacher as designer, implementer, and evaluator of children's mathematical learning environments. *Journal of Mathematical Behavior*, 6, 295–303.

Maher, C., Landis, J., & Alston, A. (1987). An analysis of changes in teaching styles of participants in a teacher development project. *Proceedings of the eleventh annual meeting of PME-NA*, pp. 218–223.

Noddings, N. (1990). Constructivism in mathematics education. In R. B. Davis & C. A. Maher (Eds.), *Constructivist views on the teaching and learning of mathematics*. Monograph 4, *Journal for Research in Mathematics Education* (pp. 4–19). Reston, VA: National Council of Teachers of Mathematics.

Peterson, P., Wilkinson, L., Spinelli, F., & Swing, S. (1984). Merging on small group process. In P. Peterson, L. Wilkinson, & M. Hallinan (Eds.), *The social context of instruction* (pp. 125–151). Orlando, FL: Academic Press.

Schoenfeld, A. (Ed.). (1987). *Cognitive science and mathematics*. New York: Academic Press.

Snyder, L. S., & Silverstein, J. (1988). Pragmatics and child language disorders. In R. L. Schifelbusch & L. L. Lloyd (Eds.), *Language perspectives* (pp. 189-222). Austin, TX: Pro-Ed.

Steffe, L. P. (1990). On the knowledge of mathematics teachers. In R. B. Davis & C. A. Maher (Eds.), *Constructivist views on the teaching and learning of mathematics* (pp. 233-254).

Webb, N. (1984). Verbal interaction and learning in peer-directed groups. *Theory into Practice 24*(1), 32-37.

Vygotsky, L. S. (1962). *Thought and language.* Cambridge, MA: MIT Press.

Wertsch, J. V. (1985). *Vygotsky and the social formation of the mind.* Cambridge, MA: Harvard University Press.

Wilkinson, L., & Calculator, S. (1982). Requests and responses in peer-directed reading groups. *American Education Research Journal, 19*(1), 107-122.

Wilkinson, L., & Spinelli, F. (1983). Using requests effectively in peer-directed instructional groups. *American Educational Research Journal, 20*(4), 479-501.

Yackel, E. (1987). A year in the life of a second grade class: A small group perspective. *Proceedings of PME-XI, 3*, 208-214.

Appendix 9A: List of Tasks

A. How many of the small blocks will we need to make a big block? Make a drawing of your answer.

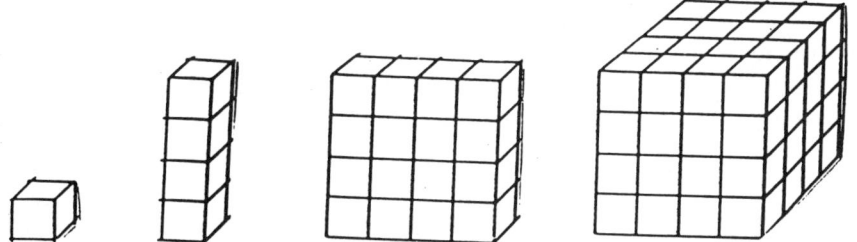

B. Erasers cost 3 cents each, pencils cost 5 cents each, and notepads cost 11 cents each. Chris spent exactly 17 cents. What did Chris buy?

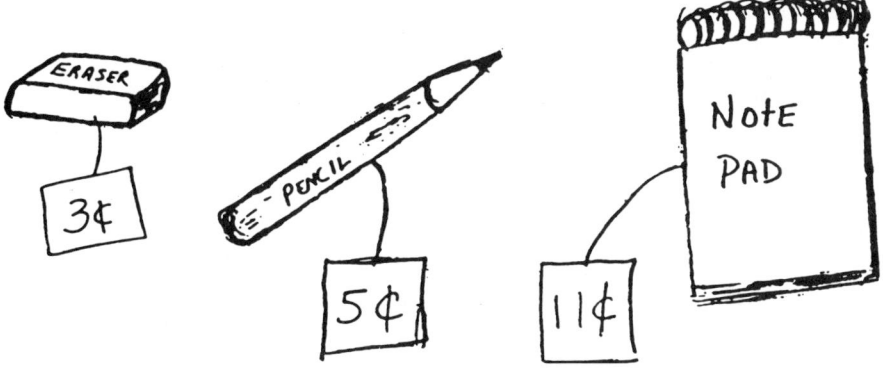

C. Mrs. McGlade ordered 7 pencils. She also ordered some erasers. She lost 6 erasers. Now the number of erasers is the same as the number of pencils. What was her order?

STORE ORDER

NUMBER OF PENCILS:____   **NUMBER OF ERASERS:**____

D. Mrs. Marinaro ordered 7 pencils. She also ordered some erasers. The number of pencils that she ordered is 5 less than the number of erasers. What was her order?

STORE ORDER

NUMBER OF PENCILS:____   **NUMBER OF ERASERS:**____

E. Mr. Rica ordered 13 erasers. He also ordered some pencils. He found 6 pencils from last year. Now the number of pencils is the same as the number of erasers. What was his order?

STORE ORDER

NUMBER OF PENCILS:____   **NUMBER OF ERASERS:**____

F. Mrs. Zemel ordered 12 erasers. She also ordered some pencils. The number of erasers that she ordered is 5 more than the number of pencils. What was her order?

STORE ORDER

NUMBER OF PENCILS:____   **NUMBER OF ERASERS:**____

Chapter Nine • Students' Disagreements   165

G. Mrs. Bender ordered 8 pencils. She also ordered some erasers. She lost 6 erasers. Now there is one eraser for each pencil. What was her order?

STORE ORDER

NUMBER OF PENCILS:_____   **NUMBER OF ERASERS:**___

H. Mrs. Brunette ordered 8 pencils. She also ordered some erasers. She ordered one eraser for each pencil and she also ordered 5 extra pencils. What was her order?

STORE ORDER

NUMBER OF PENCILS:_____   **NUMBER OF ERASERS:**___

I. Mrs. Ciechacki ordered 14 erasers. She also ordered some pencils. She found 6 pencils from last year. Now there is one pencil for each eraser. What was her order?

STORE ORDER

NUMBER OF PENCILS:_____   **NUMBER OF ERASERS:**___

J. Mr. Cliff ordered 13 erasers. He also ordered some pencils. He ordered one eraser for each pencil and he also ordered 5 extra erasers. What was his order?

STORE ORDER

NUMBER OF PENCILS:_____   **NUMBER OF ERASERS:**___

Appendix 9B: List of Interview and Sociometric Questions

Student Interview

List of Interview and Sociometric Questions
Student Interview

Present the child with large name cards of the children in his/her math group. Explain to the child that he/she would be helping you a great deal by answering some questions about him/herself and the other children in his/her math group.

1. a. Do you know these kids?
   b. Could you point to the name of the classmate that you like best?
   c. Who do you like next best?

2. a. Do you remember when you worked with these kids?
   b. How did you like working with this group?
   c. Why did you like/dislike working with this group?
   d. Would you like to work in a group again?
   e. Why or why not?

3. Let's look at the cards again.

   a. This time I want you to point to the name of the classmate who is the best math student in the group.
      Does this person get the highest grades?
      Is this the person the other kids go to when they can't do a problem?
   b. Who is the next best at math?

4. a. Have you even worked in groups like this before?
   b. How were these other groups the same or different from this math group?
   c. How long have you known the other children in this group?
   d. Were you comfortable working with this group of kids?

5. Let's look at the cards one more time.

   a. Now, when the teacher is not with your math group and some of the kids in your math group stop working on math problems that you are supposed to finish, who is the best person at getting the group back to work on the math problems?
   b. Who is the next best at getting the group back to work on the math problems?

6. Was anyone in your group in charge or "the leader"?

a. If child answered yes, ask, "Who was the leader?"
   1. If child says himself/herself ask the questions in #7.
   2. If child says another child in the group ask the questions in #8.
b. If the child answered no, ask, "Have you ever been in a group with a leader?"
   1. If the child says yes, ask the questions in #8.
   2. If the child says no, politely end the interview.

7. a. How did you like leading the group?
   b. Why do you think that you were the leader?
   c. Have you ever worked in a group where you were not the leader?
   d. What are the rules for being a leader?
   e. Do the children in the group who are not leaders have any responsibilities?

8. a. What do you think are the rules for being a leader?
   b. Do the children in the group who are not leaders have any responsibilities?
   c. Have you ever been a leader?
      If so, how did you like being the leader?

If not, would you like being the person who leads the group? Also, how did you feel about the leader?

Thank you very much, you've been very helpful!

Appendix 9C: The Cube Problem

Transcript: Group 5C—Samantha, Micky, Dava, Steve

	1. *Adult*: Here's the question, how many of the little tiny blocks make up the big block? Now you have to work on it in your groups and you can use the blocks.
2. All students use the concrete materials given them.	2. *Samantha*: [Grabs one of the cubes and a handful of units. Steve takes a cube. Micky takes a cube.]
3. Is Dava's model of the cube surface area or volume?	3. *Dava*: [to adult] Can we take the block and trace it on each of the sides? Can we put the blocks on each of the sides? [Steve begins to trace cube onto paper.]
	4. *Adult*: You can do whatever you want.
5. Samantha shares a strategy of solution.	5. *Samantha*: Let's trace this block [cube].

6. Dava disagrees, proposing different strategy.	6. *Dava*: Noooo! This is how we're gonna do it [picks up a the cube that Samantha had].
	7. *Samantha*: Well, I can do it the way I want.
9. Samantha attempts to resolve this situation by suggesting that different strategies are heard.	9. *Samantha*: [stands up and interrupts] Let's decide a way we all want to do it. [Samantha takes Micky's cube.]
10. Focusing on the sides Dava might be considering surface area.	10. *Dava*: Would you like to do it by tracing the sides. Or would you rather do it like this . . . [takes cube and units and begins building a wall of units around the cube]?
	11. *Samantha*: [Jumps in.]
	12. *Micky*: Give me that.
13. Samantha's choice of the term *fill* seems to indicate that she is thinking in terms of volume.	13. *Samantha*: Wait a second. Trace the box and then you'd be able to fill the box and figure out the problem.
14. Steve agrees with a classmate's strategy.	14. *Steve*: Yeah! Wouldn't that be easier?
15. Samantha initiates a voting process to reach a solution.	15. *Samantha*: Let's vote, who wants to do it my way? [Steve raises his hand.]
	16. *Samantha*: Who wants to do it Dava's way? [Micky and Dava raise their hands.]
17. The voting strategy doesn't resolve the conflict.	17. *Dava*: It's two against two. [Micky watches Samantha and Dava.]
18. Samantha initiates an odds-and-evens strategy to settle the conflict.	18. *Samantha*: Two against two . . . so we'll have to do odds and evens. Evens is mine, odds is hers. Ready Dava . . . one, two . . . [pause] Wait a second! You can't look at my number! . . . one, two, three . . . I won! We're doing it our way.
19. Dava states her unhappiness with Samantha's method.	19. *Dava*: I don't want to trace it.
20. The group begins to use Samantha's method.	20. *Samantha*: Well Dava . . . then we do it. [Dava, Steve, Samantha trace cube's base.]
	21. *Micky*: Where's my box [cube]?
	22. *Samantha*: Wait a second . . . I'm almost done tracing. [Micky waits for a cube.]
	23. *Steve*: [to Micky] I'll give you my block.

Chapter Nine • Students' Disagreements   169

25. Samantha attempts to explain to Dava the flaw in her strategy. Samantha provides an explanation in her own words of the difference between surface area and volume.

27. Samantha repeats her concerns to Dava, referring to the block to explain.

28. Dava persists, she wants to count the boxes on the surface of the cube.

29. Samantha again disagrees with Dava's strategy.

30. Micky shares a more efficient strategy equating 4 × 4 block with 16 units, but it is disregarded by Samantha. Samantha dissuades Micky from using his strategy.

31. Steve begins Samantha's strategy but soon seems inactive again.

24. *Micky*: But you're not almost done. [Samantha gives her big block to Micky and begins to fill in her traced box with the little blocks. Micky traces the base of a cube; Steve fills in shape with units.]

25. *Samantha*: You know, Dava, why you can't do it that way . . . because you have to work around in the middle in order to get in the middle . . . you can't pick . . . if you were going to do it your way what would happen is you would be doing the outside and you wouldn't be doing the inside. What I'm saying is you have to get all the blocks.

26. *Dava*: Yeah.

27. *Samantha*: But what you were doing is this . . . you were going around it. If you do that there is no way you can fit the blocks inside [points with her finger tracing around the block].

28. *Dava*: [Dava starts to fill in shape with the units, stops, and grabs the cube.] Wait! all you gotta do is count these.

29. *Samantha*: Dava, you can't count the inside box. [Samantha points toward the center of the cube, then continues to fill in traced box with units. Dava counts blocks on the cube. She counts 3 sides, puts the cube down and continues to fill in her traced box. Places cube alongside her cube, which is under construction.]

30. *Micky*: Here's a simple way. You just take one of these [a 4 × 4 flat block]. [He places it inside his traced box.]

31. *Samantha*: No, but we have to use the little blocks. [Micky puts back the 4 × 4 flat.] Let's get to work. [Fills in traced outline with the units, and they all begin to do it Samantha's way. Steve counts the units in his box, pushes them aside, erases, and places the cube on the paper.]

170   Part Three • What Should a Classroom Look Like?

32. Micky reaffirms Samantha's earlier statements about volume.	32. *Micky*: 8, 9, 10, 11, 12, 13, 14, 15, 16! You're not going around the outside. You're going on the inside.
	33. *Samantha*: Wait.
	34. *Dava*: O.K., we did the bottom.
35. Micky shares his answer with the group.	35. *Micky*: It's 16.
36. Samantha expresses her displeasure at being rushed.	36. *Samantha*: No, not all of us, some of us take slower time than the other.
37. Dava's mental model of the cube as solid.	37. *Dava*: O.K., now all we gotta do is make the block upper now. 17, 18, 19, 20, 21 . . .
	38. *Micky*: You gotta make the whole box? 16, 17 . . .
39. Samantha answers Micky's question.	39. *Samantha*: Yeah . . . you gotta figure out how many . . . how many . . . you'd use to make the whole block.
	40. *Micky*: 19, 20, 21, 22, 23, 24 [Building with the units. Steve begins building again.]
	41. *Samantha*: 20 count quietly . . .
	42. *Micky*: 25
43. Samantha sees a pattern and counts by fours.	43. *Samantha*: 4, 8, 16 in a row. [She starts a second layer.]
	44. *Micky*: 30, 31, 32, 33, 34, 35 . . .
	45. *Steve*: Come on, Micky, you're taking all my blocks!
	46. *Micky*: 37, 38, 39 . . .
	47. *Steve*: You took all my blocks.
48. Dava shares a more efficient method using longs with 4 blocks rather than single units. She corrects herself, replying that they're not allowed to use the longs.	48. *Dava*: I made a better way to do it . . . use these [longs with 4] units. But you can't, we're not allowed.
49. At this point, Steve ceases to exhibit any action toward a solution.	49. *Micky*: 41, 42, 43 . . . uh oh! We don't have any more ones except for these . . . oh man how many did I have? [to Steve] How many did I have just before . . . 48? [Steve puts blocks back in plate.]
	50. *Adult*: Dava, you know you can use anything here you want to solve the problem.

Samantha monitors her own construction.

51. Micky abandons Samantha's method of block by block and resumes his more efficient strategy of four pieces of wood. He rebuilds the whole figure. Dava continues her construction but with more efficient pieces. Samantha persists with her construction.

53. Micky and Dava share their answer.

57. Clearly Samantha lost count, but she connects a number symbol to her picture.
58. Micky answers Samantha's question and monitors her writing of the answer.

[Samantha takes a cube and lines it up alongside her constructed block to monitor.]
51. *Micky*: All right that's better [pushes aside the little blocks and grabs the 4 × 4 flat blocks and begins counting.] 17, 18, 19, 20, 21, 22, 23, 24, 25, 26, 27, 28, 29, 30, 31, 32, 33, 34, 35, 36, 37, 38, 39, 40, 41, 42, 43, 44, 45, 46, 47, 48, 49, 50, 51, 52, 53, 54, 55, 56, 57, 58, 59, 60, 61, 62, 63, 64! It's 64! [Dava builds builds a third layer with the flat block, counts 33–48, uses four longs for the top layer and counts.]

52. *Samantha*: [Ignores this and continues to count.] 17, 18, 19, 20, 21, 22, 23, 24, 25, 26, 27, 28 . . . 32.
53. *Dava*: I did it! The answer is 64! How do you write it? We made the block. So how do you write it do you just put down 64?
54. *Adult*: Put your answer, O.K.? And you've got your picture, right?
55. *Dava*: Make a drawing to show what you're doing.
56. *Adult*: Isn't that what you're doing?
57. *Samantha*: It was 46 . . . it was 64? [Puts the numeral inside her traced box.]
58. *Micky*: Yeah, 64 . . . don't write it down there!

59. *Samantha*: I can write it there that's the box.
60. *Dava*: All we've got to do is take the yellow sheet out.

CHAPTER TEN

Brian's Representation and Development of Mathematical Knowledge
The First Two Years

CAROLYN A. MAHER ROBERT B. DAVIS
ALICE ALSTON

This chapter is part one of a story about Brian, a student who is now in the eighth grade. It focuses on Brian's mathematical activity in grades 5 and 6. Part of a larger and more comprehensive four-year study (Maher, Davis, & Alston, 1991), this chapter describes Brian's individual and classroom mathematical activities in a K–8 school in a working-class neighborhood in the northeastern United States. This school is the site of an inservice teacher education project in mathematics that is now in its seventh year. For the past four years, Brian has frequently been videotaped while engaging in activity-based problem-solving episodes (usually with a partner)

The authors would like to express particular thanks to Pam Fisezi, the fifth-grade teacher; Janet Cariello, the sixth-grade teacher; and the students, especially Brian, Scott, and Ryan, for their cooperation, enthusiasm, and patient understanding. We would also like to acknowledge the administrative support and leadership of Anthony Richel, superintendent of the Kenilworth, New Jersey, Public School District, and Frederick Rica, principal of the Harding Elementary School.

The research reported here is based on observations of the mathematical behavior of children in classrooms that provide an environment in which meaningful mathematical explorations can and often do occur. The data from these classroom episodes are augmented by individual task-based interviews of particular children collected over time. This is possible for us because of the continuing partnership between the Kenilworth School District and the Rutgers University Center for Mathematics, Science, and Computer Education.

The present report deals with the first two years of the study of Brian. A longer report, dealing with four years of Brian's schooling, appears in the *Journal of Mathematical Behavior*, Vol. 10, No. 2 (1991).

in his mathematics classes. From these tapes we can observe how Brian has dealt with questions involving fractions, beginning with his initial introduction to fractions in 1988 and continuing up to the present time.

Some taped episodes show Brian working during mathematics lessons in school; other tapes show Brian during task-based interviews. Although it is not easy to do, it is possible to carry out very careful analyses of the videotape transcripts, from which one can learn the *fine detail* of Brian's ways of dealing with many of the key ideas related to fractions. Because we can follow these episodes over time, we can learn much about how Brian's thinking has developed (and in some ways has deteriorated) over the ensuing four years.

Background

Brian was brought to our attention in the spring of 1988 by Judith Landis, who was analyzing videotapes of children doing mathematics in a fifth-grade classroom. The first episode that we studied shows Brian using concrete materials to build an initial representation of a problem that involves fractions. Referred to as the "two-pizza lesson," the episode is described elsewhere in considerable detail (Davis & Maher, 1990; Maher & Davis, 1990; Davis, Maher, & Alston, in press). In the lesson, Brian is seen to be thoughtfully engaged in building his initial representation of the problem, using unusually concrete modeling. Not only does he use wooden Pattern Blocks to model the pizzas that are to be shared, but he takes the remarkable step of assigning the names of specific classmates to give him a concrete model of the children by whom the pizzas are to be eaten. Our interest in Brian's problem solving was immediately captured by the methodical and persistent way that he went about it, arriving ultimately at the correct answer.

Brian seemed a good candidate to follow over the years, and we have been fortunate in being able to do so. We continue to videotape classroom lessons of Brian working, sometimes with a partner, sometimes alone. Videotaped task-based interviews have been added to classroom observations to give us an opportunity to ask Brian direct questions about how he is thinking or why he has made various specific decisions. The resulting data give us a remarkable look at how Brian represents certain mathematical ideas and how his ideas, methods, and attitude change over time.

Videotapes from the Landis study of Brian working with his partner, Scott, provided us with initial data on Brian's approach to mathematics as a fifth-grader. He consistently used concrete materials or pictures to build his initial representations of problems. His thinking was primarily in terms of the meaning of the problem situation, and he made sure that he had a powerful and correct grasp of that situation (recall his representing by name the students who received the slices of pizza). Both the initial problem situation and any operations performed on it were represented by manipulable physical materials such as Pattern Blocks (see Figure 10-1).

The Landis Study

To begin our story, we report representative sample episodes of Brian's mathematical thinking as a fifth-grader. His teacher, Pam, was a second-year participant in the inservice teacher education program. As a result of this program, Pam was making a special effort to include in her lessons opportunities for children to work in small groups (usually in pairs). Characteristic of Pam's teaching that year was her practice of providing students with the freedom to think things out for themselves and (what is more unusual in American classrooms) enough *time* to think carefully. She left the choice of materials used to model problem situations to the discretion of her students. She encouraged the interaction and sharing of information and strategies, and she encouraged children to record what they built by drawing pictures and describing their solutions with numbers.

Landis's study concentrated on Pam's perceptions about the mathematical thinking of her students (Landis, 1990). To validate Pam's judgments, independent observations of the videotapes of children doing mathematics were made by Landis and two external reviewers. Differences in judgments were resolved by a team of mathematics educators. Episodes of Brian's thinking in which there was consensus among reviewers are reported here. It should be noted that in the Landis study the viewers were not trying to "see what happened, whatever that may have been," but rather, were trying to identify occurrences of behaviors from a specific, previously agreed-on list—a rather long list, which is presented in Landis and Maher (1989). In more recent studies of these same tapes, we have switched to an alternative methodology, whereby we ask observers to note *anything* of interest, because we are finding that some of the most important behaviors could not have been anticipated and prespecified. Even from the beginning, however, the question of *which* kinds of representations the children used was left open because we wanted to avoid missing anything that might function for them as some kind of representation. In addition to mental representations, which of course cannot be directly observed, representations that could be observed included representations using concrete materials (such as Pattern Blocks), representations by free-hand drawings, representations in words, and representations in terms of mathematical symbols. One can also identify what might be called "representations by imagined scenarios," as, for example, when a student said that one-third must be more than one-fourth because you were sharing with fewer other people (two others instead of three others), so your share would have to be larger. There is even one more kind of representation, which we might call "representation by choreography," where a person's gestures may make clear some action of sharing, cutting, discarding, or something else of that sort. Brian used this quite often, and one could often follow his argument by carefully watching his gestures.

Brian's Classroom Mathematics in Grade 5

In our observations of the videotapes of Brian in grade 5, we see him working with his partner, Scott. The following early episode was identified within the Landis study as an example in which the boys worked together and were thoughtful in their questioning of whether the answer that they had constructed made sense (this was one of the behaviors on the checklist of "things to look for").

Questioning One's Solution

The children were given the following problem on a worksheet:

Problem 1

Thirty-four children came in from lunch recess, and Mr. Rica, the principal, gave each one of them 1/5 of a candy bar. Name the IMPROPER FRACTION and MIXED NUMERAL that would stand for the amount of candy they got altogether. Model the amount of candy and draw your model.

Their approach to solving the problem included the following discussion:

Scott: 34/5

Brian: Six and . . .

Scott: No, seven.

Brian: No, six.

Scott: Seven.

Brian: Seven is thirty-five.

Scott: Six is thirty . . . right, am I right? Six is thirty.

Brian: Yeah, but we have to get to 35. Yeah, O.K. . . . Yeah, you're right.

Scott: Seven and . . .

Brian: Seven and . . . seven and one . . . no, seven and 4/5.

Scott: No, wait . . . wait, it is six.

Brian: See.

Scott: Six and 3/4 . . . 3/5. Six and 4/5.

Using Concrete Materials to Build Models

Other episodes were identified as examples of lessons in which the boys worked together, using the concrete materials that were available to solve the problems (another item on the checklist of things to look for). For some of the activities Brian used Pattern Blocks to build his model. One problem from these episodes is the following:

Problem 2

When the kindergarten children came in from lunch recess, Mr. Rica gave each one of them 1/4 of a candy bar. Five children came in. Model the amount of candy they have altogether.

Scott: This one's hard . . . no, this one's easy . . . [building a unit of four yellow hexagons] but you have to draw two cakes and there's no room.

Brian: Try smaller pieces. How are we going to do this?

FIGURE 10-1 • *Pattern Blocks*

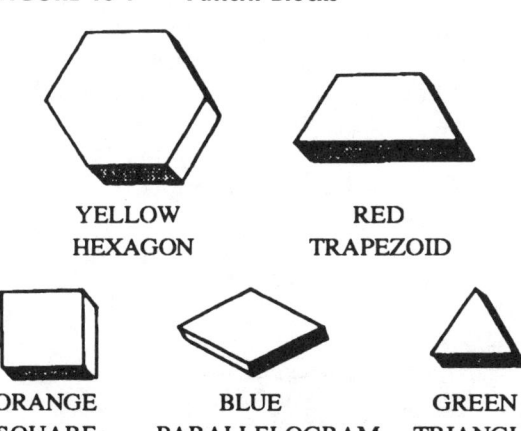

YELLOW HEXAGON
RED TRAPEZOID
ORANGE SQUARE
BLUE PARALLELOGRAM
GREEN TRIANGLE

When the boys turned to their teacher to request a sheet of blank paper on which to record their model, she suggested that they use the orange squares rather than the hexagons.

Brian: You have to shade in all of them [referring to the unit of orange squares that he built].

Scott: No.

Brian: See how there's five and five children.

Scott: There's only four.

Brian: No.

Scott: There's only supposed to be four, Brian.

Brian: Oh yeah.

Brian modified his construction to represent two units each made up of four squares instead of five. This is indicated by the darkened line between the fourth and fifth square of his drawing of eight squares in Figure 10-2.

Brian: Now we just shade in five of them. Scott, it's 1 and 1/4.

Scott: [reading from the worksheet] Name this as an improper fraction.

Brian: As a mixed number . . . 1 and 1/4. [Brian wrote 5/4 on his worksheet after shading the region indicated in his drawing.]

Relating One Representation to Another

Other episodes also show Brian relating one kind of representation to some representation of a different sort. In one instance, he was observed drawing and shading diagrams, and then counting, in order to get from a picture representation to a representation in terms of mathematical symbols. The problem was as follows:

FIGURE 10-2 • *Brian's drawing for Problem 2*

Problem 3

Eleven children came in. Model the amount of candy they got. How many fourths is this? How many whole bars and part of a bar is this?

Brian: [He drew a picture representing three units, each built of four squares and as he shaded it said aloud] 4, 8, 11. [Pointing to his drawing, he said] O.K. . . . 1/4, 2/4, 3/4, 4/4, 5/4, 6/4, 7/4, 8/4, 9/4, 10/4, 11/4. . . . 2 and 3/4.

Figure 10-3 shows Brian's drawing of Problem 3.

Building a Representation of the Problem Situation

One further example illustrates Brian's use of building a pictorial representation to solve a problem. Notice that Brian represents the problem situation so that it is related to the actual story.

In problems 4 and 5, the students were asked to model situations as improper fractions and mixed numbers.

Problem 4

Mrs. A. brought in super-duper, giant-sized chocolate chip cookies. She broke each cookie into thirds and gave 1/3 of a cookie to each of 11 girls. Model what Mrs. A. gave away. Draw a picture of your model. How many whole cookies and part of a cookie did Mrs. A. give away?

In this problem it seems to have been intended that *model* meant "make a model with some kind of physical materials"; otherwise the instruction "Draw a picture of your model" is difficult to interpret.

FIGURE 10-3 • *Brian's drawing of Problem 3*

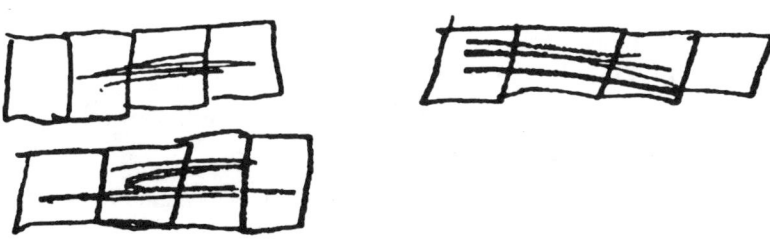

Brian and his partner worked together, choosing the yellow hexagon as their basic artistic unit, but they may not both have meant the same thing by this choice. After rereading the problem, Brian drew a picture of four "hexagon cookies" and divided each into thirds. Scott, meanwhile, was busily drawing hexagons on his paper. Brian looked at Scott's drawings and asked:

Brian: How many of these things do we have to draw?

Scott: Eleven.

Brian erased his first drawing of four hexagons, each partitioned into three parallelograms, and began tracing eleven hexagons on his paper. Something interesting is going on here. We have three clues as to what it might be (and a fourth clue will appear later in this episode). First, if you have drawn four hexagons, and decide that you need not four, but eleven, it hardly makes sense to *erase* the four that you have already drawn, and to start over. Second, why did Brian change his mind and switch from the goal of having four hexagons to the new goal of having eleven? Finally, where did the number 11—and particularly the number 4—come from? One hypothesis would explain all three mysteries: Brian started out using a hexagon to represent a cookie. On other occasions Brian has demonstrated that he knows the relation 3 times 4 is 12 and uses it in deciding how many wholes go with how many thirds or fourths. Hence it would be in character for Brian to reason: "Each girl gets one third. There are 11 girls. So I need 11 thirds. But 4 cookies would give me 12 thirds, so I need 4 cookies." If in fact he reasoned this way, then it made sense to draw four hexagons.

Now, why did he change? Perhaps he saw Scott drawing *eleven* hexagons and turned his attention to the number 11 (which, in fact, he explicitly asked Scott about). Clearly there weren't eleven cookies; the "11" referred to eleven *girls*. Brian now erased his picture of *cookies* and made a new picture, this time a picture of *girls*. It did not occur to him that a picture of a cookie can just as well be a picture of a girl, if in fact both pictures are really drawings of hexagons.

But even with Brian's new picture, Scott was not satisfied,. He criticized Brian's drawing because some of the hexagons shared a common edge.

Scott: You're doing it wrong.

Brian: They're [drawing dark lines between the hexagons] not really together. Now what do we do?

Scott: Now let's see . . . This [indicating the eleven hexagons] represents the eleven girls and they each get 1/3 of a cookie.

At this point it seems clear that Scott is not using a hexagon to represent a cookie but, rather, to represent a girl. It is less clear what each hexagon means for Brian. We would hypothesize that Brian started out with each hexagon meaning one cookie. Seeing Scott's drawing of *eleven* hexagons, Brian switched and for a moment construed (as Scott was doing) that each hexagon represented one girl. But Brian now appears to have switched back and to see in front of him eleven cookies, from each of which only one-third is to be eaten. Fantastically wasteful, is it not? Brian is seeing 22 thirds of cookies left unaccounted for.

It is here that the fourth clue appears, for Brian is disturbed by this waste. Scott is not. Of course Scott isn't disturbed; for him there is no waste. He is looking at 11 pictures of girls.

Brian: One-third? What happens to the rest?

Scott indicated that he didn't care about what happened to the leftover cookies. Brian then picked up a blue parallelogram and used it to partition each of the eleven hexagons into thirds. He then placed a mark in one of the thirds of each hexagon, as indicated in Figure 10-4. After he finished marking a third in each of the hexagons, he began counting aloud, "One . . . two . . . three," and, covering three hexagons with his hand, he said, "A full cookie." Brian then repeated this action, covering a second group of three hexagons, and added, "See, we now have two full cookies." He covered a third group of three hexagons and said: "One . . . two . . . three . . . a full cookie." He then asked: "So how many? That's three cookies." Scott replied, "and 2/3 more."

FIGURE 10-4 • *Brian's solution to Problem 4*

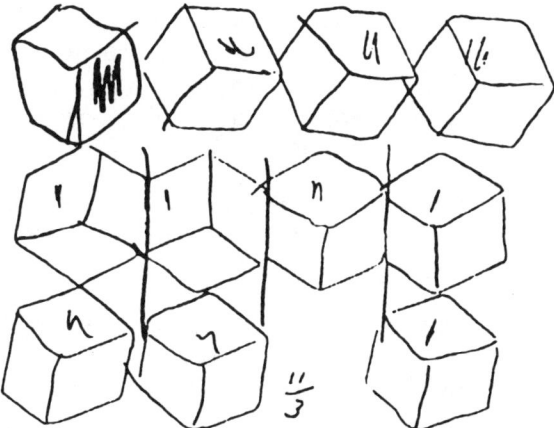

182 Part Four • The Reality of Students

Brian wrote 3 and 2/3 as the numerical solution to the problem. The boys then moved on to the next problem.

So — we have four mysteries, and one hypothesis that explains them all. Do we know that we are correct in our assessment of what the boys had in mind? Of course not. Analyzing videotaped episodes is like anything else in science; on the one hand there are the data — which in many sciences means little needles pointing to numbers on dials — and on the other hand there is the theoretical conceptualization of what is going on, that hopefully makes sense out of the numbers to which the needles are pointing. We would argue that our hypothesis does explain at least four things:

1. Why Brian drew *four* hexagons, and not some other number
2. Why he changed his goal, abandoning his original aim of drawing four and adopting the new goal of drawing eleven
3. Why, with four hexagons already drawn, when he decided that he needed eleven, he began by *erasing* the four that he had already drawn
4. Why Brian, but not Scott, was concerned about "what happens to all the other pieces of the cookies"

There is at least one more reason that we are disposed to accept our present hypothesis. Time and again (see Chapter 1, this volume, or Davis, 1984) we see people making errors at the same critical point in their thinking — specifically, *the point where they try to map some input data into the variable slots in their mental representation*. In Chapter 1 we saw a student draw a line segment to represent "a three-mile race." He then needed to represent the statement "Mary ran 3/4 of a mile in the first 5 minutes." He did so by marking a point three-quarters of the way along the line segment, forgetting that the segment had been intended to represent the *race*, not to represent a mile.

Other researchers have often reported errors occurring at this same stage in thinking. In an important analysis of the so-called students-and-professors problem, Rosnick and Clement (1980) report the case of Peter, a university student in physics and calculus who was working on a variant of the original students-and-professors problem that used people in China and in England, rather than "students" and "professors," and was stated as follows:

> There are 8 times as many people in China as there are in England. Using E for the number of people in England, and C for the number of people in China, write an equation expressing this fact.

Peter, like so many other students, became confused about how to use the letters C and E, with the result that he produced a nonsensical equation. Later, when using the *correct* equation, $8E = C$, he continued to be overwhelmed by his earlier wrong interpretation and tried to reconcile his idea with the evidence of his eyes. The following dialogue ensued:

Interviewer: So what does the C mean there?

Peter: C means the English person . . . uh, England itself.

As Rosnick and Clement remark, "Peter's misconceptions are so resilient that he is willing to associate the letter E with China and the letter C with England rather than change his internal conceptualization" (Rosnick & Clement, 1980, pp. 11–22).

Or, to put the matter into our present language, Peter keeps his representation but changes the way he maps input data into the variable slots in that representation—just as, we believe, Brian is prepared to change the meaning of the hexagons from "cookies" to "girls" and then back to "cookies"

Using a Guess-and-Check Strategy

In the Landis study, the reviewers noted Brian's use of a guess-and-check strategy to solve Problem 5. In this episode Brian and his partner were trying different Pattern Blocks in order to come up with something that would help them to model fourths. Brian's final choice for a unit was two yellow hexagons.

Problem 5

Mrs. A. broke the cookies into fourths, 1/4 for each boy. She gave away 14 fourths. Model what Mrs. A. gave away. Draw a picture of your model. How many whole cookies and part of a cookie did Mrs. A. give away?

Here is how the two boys worked on the problem:

Brian: What can go into fourths? . . . Reds! . . . No.

Scott: Green [triangles].

Brian: Green is six. [Six triangles cover the hexagon; see Figure 10-5.]

FIGURE 10-5 • *Equivalence of one hexagon to six triangles*

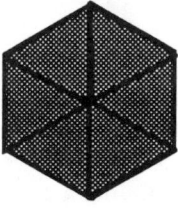

184 Part Four • The Reality of Students

FIGURE 10-6 • *Equivalence of one parallelogram to two triangles*

Scott: What does one blue [parallelogram] equal? . . . Two greens [triangles; see Figure 10-6].

Brian: Yeah!

Scott: We might have something there . . . That's wrong.

Brian: What would two blues equal to?

Scott: Two blues equal four [triangles; see Figure 10-7].

Brian picked up a red trapezoid, to which Scott responded, "No." Brian continued and estimated the number of green triangles required to cover a red trapezoid.

Brian: Greens . . . we could use to equal . . . wait! . . . what about? How many greens equal red?

Scott: How many greens equal red?

Brian: Four. [Brian then manipulated some green Pattern Blocks, discarded them, and chose some blue ones.]

Scott: It doesn't fit.

Brian then chose to work with red trapezoids and blue parallelograms and then discarded them.

FIGURE 10-7 • *Equivalence of two parallelograms to four triangles*

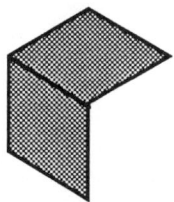 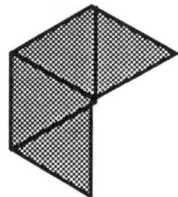

FIGURE 10-8 • *Two green triangles placed on top of trapezoid*

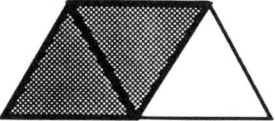

FIGURE 10-9 • *Brian's drawing of one unit partitioned into fourths*

Brian: Take two of these [indicating the red trapezoids] . . . Scott, I might be right . . . green, green, green, green. [Brian then placed two green triangles on top of a red trapezoid as indicated in Figure 10-8.]

Scott: Possibility.

Brian continued for some time arranging the blocks in different combinations before deciding to use four red trapezoids or, equivalently, two yellow hexagons, for his unit. Then he said aloud, "I've got it." On his paper he drew 14 units. Figure 10-9 illustrates his choice of one unit.

More detailed analyses of some episodes of Brian doing mathematics in grade 5 appear elsewhere (Davis & Maher, 1990; Maher & Davis, 1990; Davis, Maher, & Alston, 1991). The examples used for this chapter were selected because they illustrate particular mathematical behaviors identified by Brian's fifth-grade teacher and by external reviewers of the videotape segments, working from a preplanned list of "things to look for."

Interviews with Brian in Grade 6

An Interview with Brian before Beginning a Unit on Fractions (February 15, 1988)

The following year we decided to interview Brian before and after he began his sixth-grade work in the classroom on fractions.

Interviewer: I'm going to ask you some more questions about math and you can figure out the answers any way you like. You can use all these things also. Have you used all these things?

Brian: Yeah.

Interviewer: The Cuisenaire rods. How did you use them?

Brian: Ah, . . . like we made trains and like, say, we took an orange . . . and we take other colors and try to fit them in.

Interviewer: Yeah . . . then I know you used the Pattern Blocks . . . How did you use them? . . . Do you remember? . . . How did you use the Pattern Blocks?

We see in the videotaped episodes of Brian's problem solving in grade 5 a rather comfortable use of Pattern Blocks to model solutions to problems. It is interesting to note Brian's comment about the extent of their use.

Brian: Well, I don't think we've used these [this year] . . . but last year . . . We just tried building figures and she gave us figures to fill them in. Like . . . she said there are a certain amount of ways you can fill them in and we had to figure out how many ways with just one color we could stick it in. [Notice that, because of the videotapes, we know that this is a very misleading description of how the students had used Pattern Blocks the previous year. It may also be an indication of a problem: As Brian "filled in figures using only one color," was he fully aware of this as a definition of fractional parts? Or was he just "covering wooden blocks with some other wooden blocks"?]

Interviewer: Yeah, that's right and [you] did problems with them too, as I remember . . . and then the geoboards, did you use them?

Brian: Yeah.

Interviewer: Where?

Brian: Um. We used them to find the area of figures.

Interviewer: Yeah, and you've used the chips also?

Brian: Uhm. Uhm.

Interviewer: Have you used those for anything?

Brian: Um. I think once this year, But . . .

Interviewer: In the past? Other times?

Brian: Yeah. Last year we used them a lot.

Interviewer: Um. O.K. . . . but we can use any of them now. Now I want to ask you a question. Is 1/3 bigger than 1/4 or is it smaller?

Brian: Bigger?

Interviewer: How do you know?

Brian: I . . . because if you take 1/3, take this [reaching into the pattern blocks and placing one blue parallelogram on top of a yellow hexagon] would be one part of it. and if you wanted to fill this in with 4 . . . Um . . . it would be smaller so . . .

Brian held one yellow hexagon in his hand as he talked. Figure 10-10 shows his choice of unit and his representation of one-third.

FIGURE 10-10 • *Brian's models of one unit and one-third of the unit*

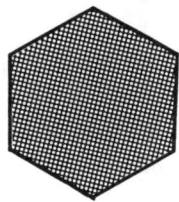

One Unit

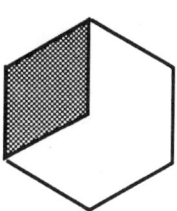

One Third of One Unit

FIGURE 10-11 • *Brian's model of three thirds*

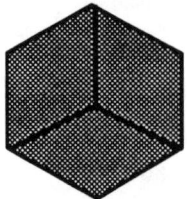

Interviewer: Could you show me that? Suppose you wanted to explain this to a little child who didn't know anything about fractions. How could you explain that?

Brian: Fill this in [indicating a hexagon] and here's three parts of it [indicating three blue parallelograms, as in Figure 10-11].

Brian: . . . and I don't think anything goes into 4 . . . really, . . . six . . .

Brian then placed six green triangles on the hexagon, as indicated in Figure 10-12.

Brian reasoned that if it was possible to partition the single hexagon into fourths, one of the four sections would be larger than 1/6 (illustrated in Figure 10-12) and smaller than 1/3 (illustrated in Figure 10-11). He pointed out that the third would be larger than the fourth because fewer larger pieces are being used to cover the unit. Brian held up a blue parallelogram as indicated in Figure 10-13 and continued.

Brian: But if you did do four . . . If you take out one [blue] and one of the things that makes four. This would be bigger [blue] because you're using less . . . Um . . .

FIGURE 10-12 • *Brian's model of sixths*

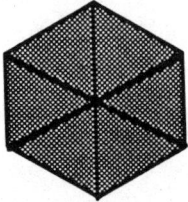

FIGURE 10-13 • *Brian's representation of one-third*

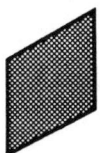

The interviewer then prompted him to build a model to demonstrate his argument with objects. Brian then turned to the Cuisenaire rods and studied them, but chose not to use them, indicating that he was unable to find one rod that could be matched with trains of both three and four rods of the same length. He then tried to use a geoboard to build a model. He began by enclosing four square units. This model was also discarded because he was unable to partition the region into three equal parts. The interviewer then offered the suggestion that Brian reconsider using the Pattern Blocks. She suggested that he consider using more than one block to represent a unit or part of a unit. After this suggestion, Brian, with some initial difficulty but with encouragement from the interviewer, built a unit consisting of two yellow hexagons as indicated in Figure 10-14.

Brian then covered this with four red trapezoids, each of which he identified as 1/4 of the unit (Figure 10-15).

Finally, he covered another pair of hexagons with six blue parallelograms, as indicated in Figure 10-16.

He then suggested that two of the blue parallelograms together would represent 1/3 of the unit as indicated in Figure 10-17, and that they, together, were larger than the single red trapezoid, as indicated in Figure 10-18.

The interviewer then asked Brian to draw a picture of what he had built. She then continued:

FIGURE 10-14 • *Brian's model of 1*

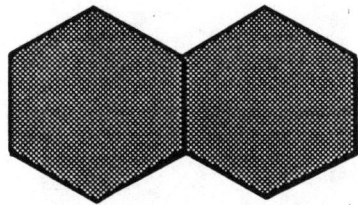

FIGURE 10-15 • *Brian's model of 4/4*

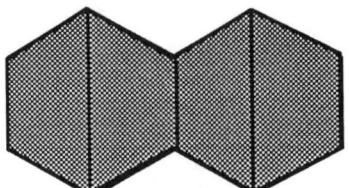

FIGURE 10-16 • *Brian's model of 6/6*

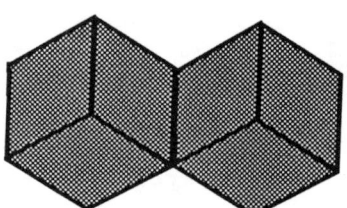

Interviewer: Now you told me this was one-third [pointing to the two blue parallelograms]. Can you show me two-thirds?

Brian: It would be this [pointing to a part of the drawing].

Interviewer: Can you show me? This is the 1/3 [indicating the outlined portion of the drawing.] . . . I don't know what 2/3 is?

Brian: This would be the 1/3 [placing two blue parallelograms on the unit he had built of two hexagons] and if you wanted 2/3 . . . it'd be added on to that [placing 2 more blue parallelograms on his unit to indicate two-thirds as in Figure 10-19].

Interviewer: That would be 2/3? Um . . . What about 4/3? . . . Can you show me 4/3?

Brian: 4/3? You would add these two. [Brian placed two more blue blocks on top of the remaining space to completely cover his original unit, as indicated in Figure 10-20.]

Brian: That would be three thirds . . . You have to add another one [hexagon] to make 4 . . .

FIGURE 10-17 • *Brian's model of 1/3*

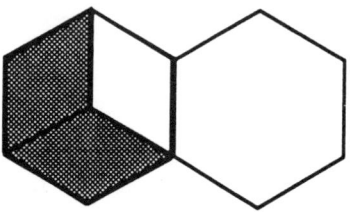

FIGURE 10-18 • *Brian's model of 1/4*

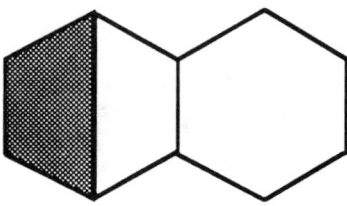

FIGURE 10-19 • *Brian's model of 2/3*

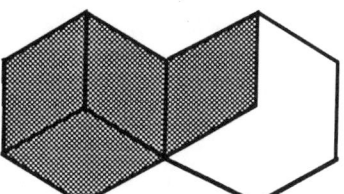

FIGURE 10-20 • *Brian's model of 3/3*

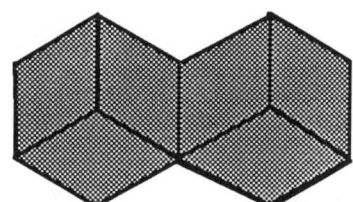

Brian then brought in another yellow hexagon to add to the original two, and then brought in two more blue parallelograms to add to the original six.

Brian: That would be 4/3 [Brian counted the blue blocks while pointing out the four sets of two, as indicated in Figure 10-21].

Interviewer: 4/3? So this is your whole one now? [She pointed to the three hexagons.] . . . Your whole unit?

Brian: Yeah.

Brian's tone of voice makes it clear that he is satisfied to regard the *three* hexagons, taken together, as his unit. (This, incidentally, is one more instance of the kind of "mapping" error discussed earlier.) The mapping had been set up to have *two* hexagons, considered together as a pair, correspond to "one unit." Now, apparently without realizing that he is doing it, Brian changes the correspondence so that *three* hexagons map into the concept of "unit." Time and again one finds that errors are made at precisely this critical stage of processing—setting up the correct correspondence between input data and items in the mental representation.

FIGURE 10-21 • *Brian's model of 4/3*

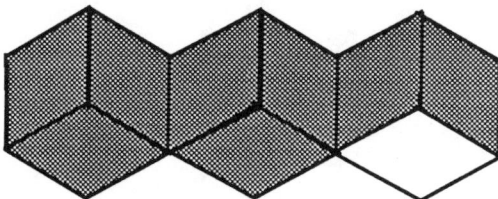

192 Part Four • The Reality of Students

Brian, when questioned about the number of pieces necessary to cover the new unit of three hexagons, admitted that each blue parallelogram was now 1/9 of the unit and that two blue parallelograms together no longer represented 1/3. The interviewer then suggested that he think of the unit as a candy bar and asked him to show her what he might use to represent a candy bar. Brian again built the "candy bar" using two hexagons and indicated his models of 1/3, 2/3, and 3/3 of the "candy bar" (Figure 10-22).

Brian was now asked to show 4/3 of the "candy bar." He responded, with some uneasiness, as follows:

Interviewer: Now suppose you were going to show me 4/3 of the candy bar?

Brian: 4/3? . . . I'm not sure. I think you would . . . I'm not sure . . . Like here . . . [adds a third hexagon, as in Figure 10-23] It would be this? . . . I'm not sure . . . I don't remember.

Later in the interview there was occasion to return to Brian's drawings that illustrated the models he built of 1/3 and 1/4. Notice that Brian's reasoning is in terms of the model he built earlier. He now justifies his number solution of one-twelfth by saying one-third is "about a triangle bigger" than 1/4.

Interviewer: O.K., now I want to go back to our 1/4 and 1/3 which you were working on right here. If one of those two fractions is larger . . . you said one was bigger than the other?

FIGURE 10-22 • *Brian's model of 1/3, 2/3, and 3/3*

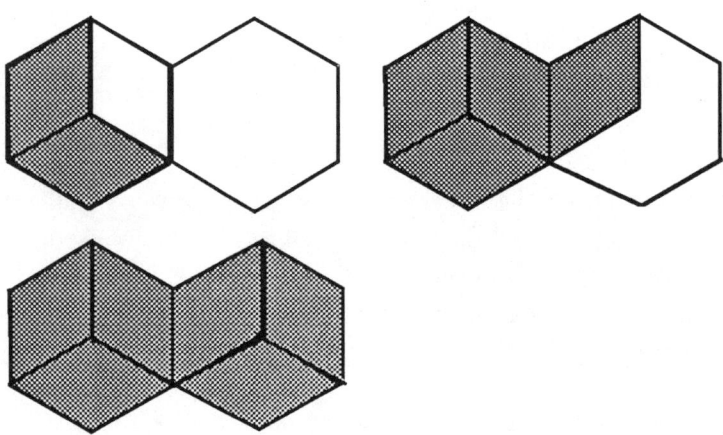

FIGURE 10-23 • *Brian's attempt to model 4/3*

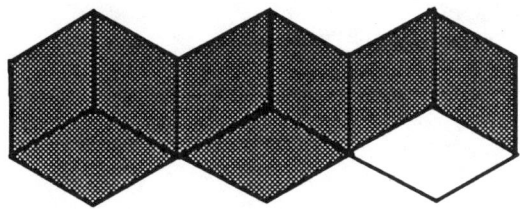

Brian: Yeah.

Interviewer: Which did you say was bigger?

Brian: The 1/3.

Interviewer: How much bigger is it?

Brian: Um. Um . . . about a triangle.

Interviewer: Show me.

Brian: Here's the 1/3 and here's the 1/4. It goes over and there'd be a triangle left [see Figure 10-24].

Brian built a model with the Pattern Blocks, placing a red trapezoid on top of two blue parallelograms, then putting a green triangle next to the red trapezoid.

FIGURE 10-24 • *Brian's use of Pattern Blocks to compare*

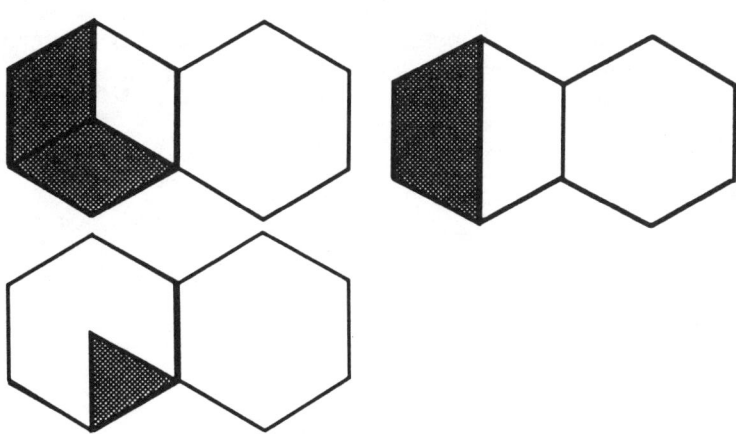

Interviewer: Oh, so there'd be a triangle left?

Brian: To fit in.

Interviewer: Yeah, so it would be a triangle bigger? How much bigger is that as a number?

Brian: Um, about . . .

Interviewer: What is the triangle as a number?

Brian: Is 1/3 and about . . .

Interviewer: How do you figure out what fraction of the candy bar— of your unit—is the green [triangle]?

Brian: Of this? (pointing to the two parallelograms which was his representation of 1/3 as in Figure 10-22].

Interviewer: No, of your whole.

Brian: About . . . 1/6?

Interviewer: Show me. How did you figure that out? What was your candy bar?

Notice how Brian is able to monitor his own reasoning when he thinks about parts of a candy bar.

Brian: Oh, no, . . . wait . . . not 1/6, like one [triangle] . . . and there's my candy bar and you have one out of 12 so it would be 1/12.

Interviewer: O.K., so how much larger, then, is 1/3 than 1/4?

Brian: Um . . . 1/12? This is one bigger [one green] and 12. Fit them in . . . , So it would be 1/12 [see Figure 10-25].

Interviewer: O.K. [setting the materials aside] Can you make up a story problem that would go with the 1/3 and 1/4? Can you think of a story problem that would go with this?

FIGURE 10-25 • *Brian's model of 1/12*

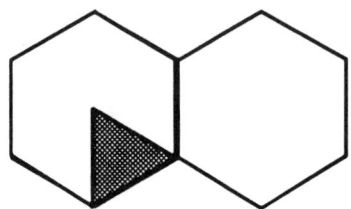

Brian: Write it down?

Interviewer: Um, yes.

Brian: All right.

Brian wrote the following problem:

Three boys were standing at a corner. One said: "Hey, I have some candy bars. Do you want some?" So he gave one of the boys 1/3 of his candy bar and the other boy got 1/4 of his candy bar. One boy said: "Hey, I only have 1/4 of the candy bar and he has 1/3 of the other. Why does he get more than me?" How much more does the boy with the 1/3 get than the boy with the 1/4?

Interviewer: How would you solve it?

Brian: Um, if somebody had, . . . wait . . . then we could draw a picture and try and get in the picture and try and see if they could stick three parts into it and four parts and really to see the difference you would stick four parts.

Interviewer: Sort of like you did?

Brian: Yeah, and put them together and see which one is bigger?

Interviewer: Could you solve it using numbers just like in arithmetic? How would you solve it just using numbers?

Brian: Um . . .

Interviewer: Could you solve that problem with just numbers, if you didn't have pictures and things?

Brian is convinced of the validity of his solution of one-twelfth based on the model he built. To justify the answer with numbers, he merely multiplies one-third and one-quarter, seemingly unaware that the structure of the arithmetic makes no sense. He clearly has not connected his concrete solution to his symbolic representation.

Brian: I would say, cause you could like . . . 1/12. Times the bottom parts and you leave the top parts the same . . . and you'd equal out to 1/12. Like the 3 . . . and if you times it times the 4 . . . just leave the top parts alone, the one. Then you'd get 12.

Interviewer: That's interesting.

Brian: Well, that's what I think. I'm not sure.

Interviewer: Is that what you think that you just times it and get the 12?

Brian: Yeah,

Interviewer: Yeah, um, . . . and because you know it has to be 1/12.

Brian: Yeah.

Interviewer: How did you know it has to be 1/12?

Brian: When I put the shapes together, um . . . , it, . . . I did two shapes and it was a triangle bigger and if I was to take the triangle and this [indicating the 2 hexagons], it take 12 triangles to fill it up, so it would be 1/12.

Interviewer: So the difference is 1/12?

Brian: Yeah.

Interviewer: Yeah, Show me the 1/4 again.

Brian: 1/4?

Interviewer: Yeah.

Brian: This [he held up a red block].

Interviewer: And show me the 1/3 again.

Brian: This [he picked up two blues].

Interviewer: O.K, . . . and so you find the difference and multiply?

Brian: Oh, . . . that's how I, . . .

Interviewer: That's how you did it here.

Brian expresses his own self-awareness about his ability to figure out a solution when given an opportunity to build a model. He also seems aware of his lack of understanding in applying arithmetic procedures.

Brian: Yeah, maybe, . . . um . . . I'm not that sure about numbers, but if I just did it with stuff like these I could figure it out.

Interviewer: Yeah, so you can figure it out by using something like that and you really believed what you found, didn't you?

Brian: Yeah.

Interviewer: Yeah. And you believed it was 1/12.

Brian: Yeah.

Interviewer: Why don't you write down the answer [Brian wrote beneath his problem], O.K., and you said that if you were going to have to do it with numbers, you're not sure. But what did you say you might do?

Brian: I timesed it. I timesed the bottom parts.

Interviewer: Could you write that down so I can remember what you said?

Brian: Like 1/3 times 1/4 equals to 1/12.

The interview continued with Brian solving another problem by correctly constructing with Pattern Blocks a model showing the relationship between a unit, 1/3 of that unit, and 1/4 of the 1/3, which he correctly identified as 1/12, as in Figure 10-26.

After solving the second problem, Brian was asked to compare the two problems. In his explanation of his symbolic solutions, he is unaware of a difference. However, he does indicate the difference when he compares the models he built.

Interviewer: O.K., now let's look at this one and this one [showing the papers]. We've done two story problems. This one went with this picture. Is that right?

Brian: Yeah.

Interviewer: And this one went with this picture.

FIGURE 10-26 • *Brian's model of 1, 1/3, and 1/4 of 1/3 of a candy bar*

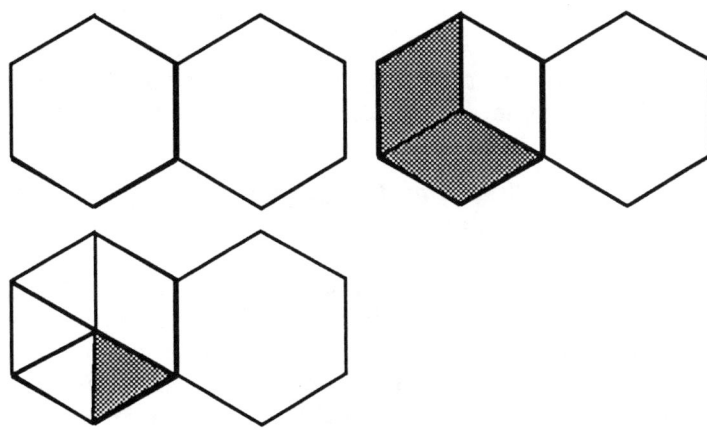

198 Part Four • The Reality of Students

Brian: Yeah.

Interviewer: How are these two problems alike and how are they different?

Brian: They are alike because they both have the 1/3 like that you have to show and the 1/4 they both came out to be 1/12.

Interviewer: How are they different? How are these two stories different?

Brian: This one [referring to the first problem]. It said how much more would he get. Then, like in here . . . and this is just comparing like [referring to the second problem]. . . . She gave him a section of it.

Interviewer: Oh, she gave him a section?

Brian: Of what she had.

Interviewer: Oh, of what she had. And what about over there?

Brian: And he . . . like a part of both candy bars to each and one got more so.

Interviewer: And so you think that was what was different?

Brian: A little.

In the final portion of the interview, Brian was asked to summarize what he had done. He held up each of his papers and described the problem in which he compared 1/3 and 1/4 by building a model with the blocks to indicate a difference of 1/12, as illustrated in Figure 10-26. Once again, when asked to explain his numerical solution, Brian referred to multiplication.

Brian: He got 1/12 more. Here is what I thought.

Brian pointed to the numerical solution that he had written out on the paper. The interviewer then asked him how he would get 1/12 from the 1/4 and the 1/3. Brian responded: "I timesed it."

An Interview with Brian after Completing a Unit on Fractions (June 7, 1989)

By now, Brian had completed classroom instruction that dealt with fractions. The intent of this second interview was to see if Brian, four months later, would respond differently to questions comparable to those that had been posed in the first interview. In the following segment, Brian was asked to compare two fractions and find the difference between them.

Interviewer: Is two-thirds larger than three-quarters? You could use any of these things to help you answer. Use your numbers . . . While you're thinking, let me know what you're thinking about.

Brian: O.K. You draw something like this [draws a rectangular shape]. Put this into three [draws vertical lines to divide the shape into three parts]. Now the same one [draws another rectangle, approximately the same size and shape]. Put it in fours. Then three-quarters, you said? So that's about three-quarters [divides the second shape into four parts]. They look about the same. No, wait. Three-quarters looks bigger.

Interviewer: You're saying three-quarters is bigger? How would you prove that to, let's say, a little kid?

Brian first reached for the bucket of Pattern Blocks and attempted to construct a model of the problem, but his choice of units did not permit him to cover both three out of four and two out of three equal parts. After several different attempts, he put aside the blocks and selected the Cuisenaire rods (Figure 10-27).

Brian: Try these.

Brian opened the box of rods, reached for two orange and two red rods, and built two equal trains, each made of one orange and one red rod as shown in Figure 10-28 to represent his unit.

FIGURE 10-27 • *Cuisenaire rod color names*

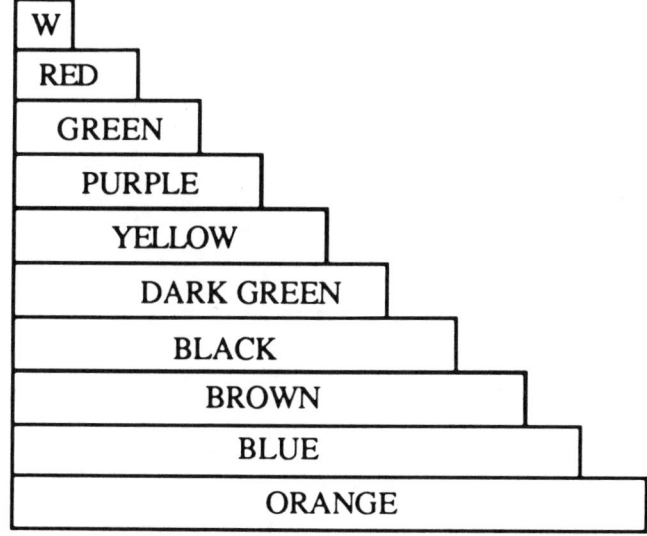

FIGURE 10-28 • *Brian's model of one unit*

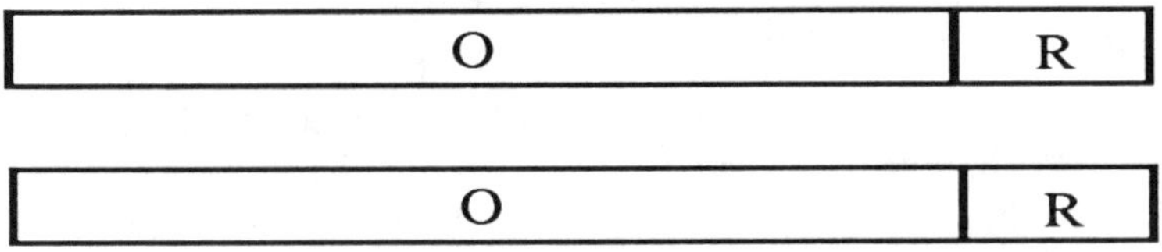

O	R

O	R

He then placed three light green rods on top of one of the trains and indicated that they would represent three-quarters, as indicated in Figure 10-29.

Brian: This would be one, see this would be three quarters.

The interviewer asked him to justify his answer.

Interviewer: How do you know that?

Brian: Cause four, four go into it.

Interviewer: Pretend I'm a real little kid and actually show me the four.

Brian: O.K.

Brian then picked up a fourth light green rod and placed it next to the other three light green to completely cover the orange and red train, as indicated in Figure 10-30. Brian then explained how he would represent 3/4.

Brian: Four—and if you wanted three-quarters, you take that away and you'd have three.

FIGURE 10-29 • *Brian's model of 3/4*

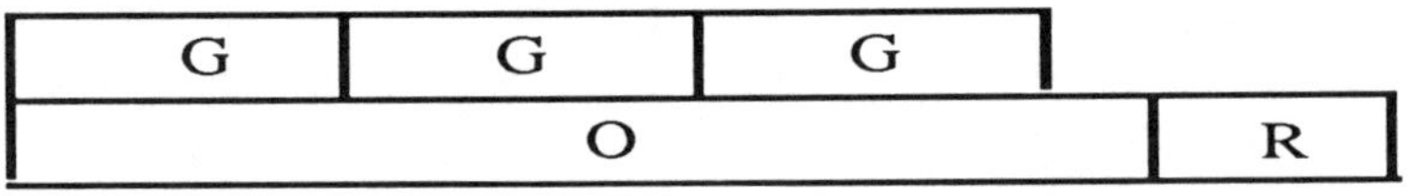

G	G	G	
O			R

FIGURE 10-30 • *Brian's model of 4/4*

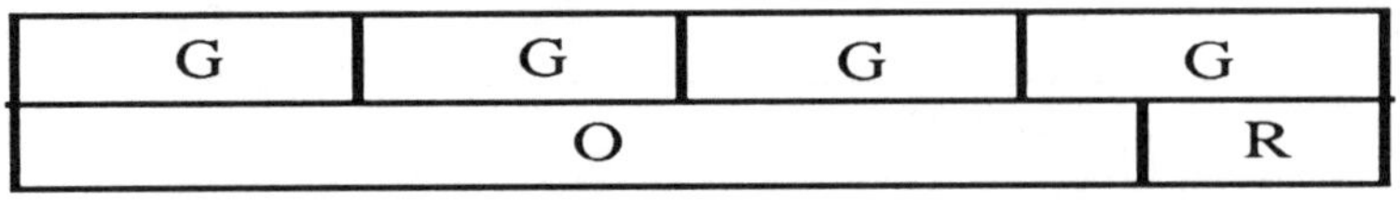

G	G	G	G
O			R

Brian then took out purple rods to place on top of the other orange and red train.

Brian: And two-thirds. This would be two-thirds. [Brian placed two purple rods on the second train as indicated in Figure 10-31, saying; "'cause three would go into it."]

Brian then picked up a third purple rod and placed it next to the others to completely cover the orange and red train, as indicated in Figure 10-32.

Brian then noted that the three light green rods taken together were longer than the two purple rods.

Brian: . . . and take that [Brian removed one of the purple rods], and it's easy to see that's [pointing to the train covered with the three light green pieces] the bigger piece of the two.

Interviewer: O.K. So now pretend I'm that little kid and explain to me again.

Brian: Well . . . Explain it . . . This would be three-fourths, . . . three-quarters, . . . and this would be two-thirds [pointing to each model in the figure above] . . . and you want to know which one's bigger. This would be the same . . .

Brian moved the small red rod away from each train, leaving the two orange ones, one with three light greens and the other with two purple rods, as indicated in Figure 10-33, saying:

Brian: Just compare these and see . . .

FIGURE 10-31 • *Brian's model of 2/3*

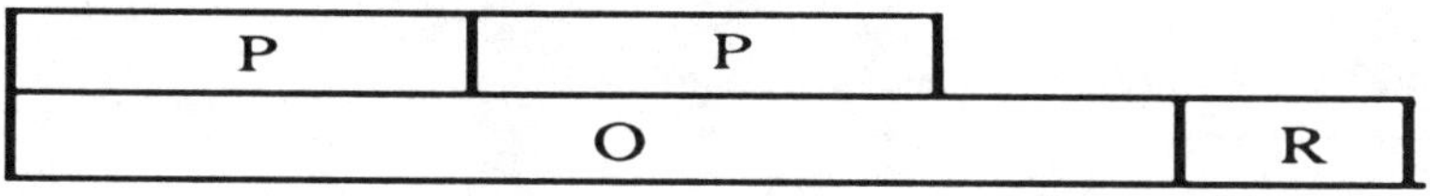

P	P	
O		R

FIGURE 10-32 • *Brian's model of 3/3*

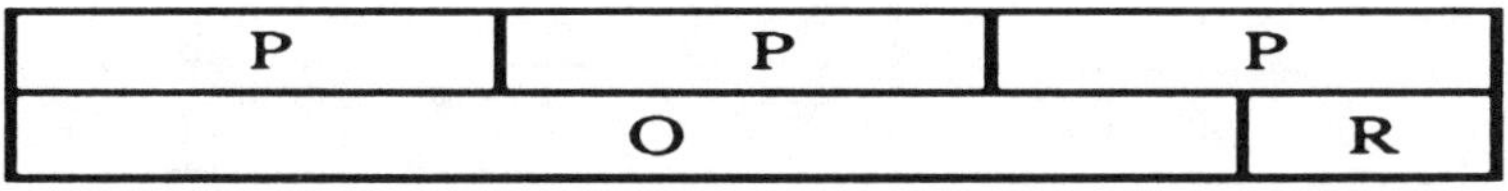

P	P	P
O		R

He then noted that the three light green rods taken together were longer than the two purple rods.

Brian picked up the small white rods and added them to each of the two models to completely cover each orange rod, as shown in Figure 10-34. He used this model as a justification for his comparison of 2/3 and 3/4.

> *Brian*: This would have one [a white rod was added to the three light greens] and this would have two [two white rods were added to the two purple rods].

The interviewer then asked Brian to write number names for what he had done, and he wrote 3/4 and 2/3. He was then asked to name the white block, which he had indicated as the difference between the two.

> *Interviewer*: Can you give me a number name for this block that tells me how much more three quarters is [than two-thirds]?
>
> *Brian*: 1/12.
>
> *Interviewer*: How would you prove that?
>
> *Brian*: 'Cause twelve of these [indicating a white rod] go into this . . . go into the whole thing.

Brian, when requested by the interviewer to show her how he would explain what he did, removed the three light green blocks from the three-quarter model and covered the orange and red train with twelve small white rods, as indicated in Figure 10-35. He said that if the two fraction models were each covered with white rods, the difference of one white rod would therefore be 1/12.

FIGURE 10-33 • *Brian's comparison of 2/3 and 3/4*

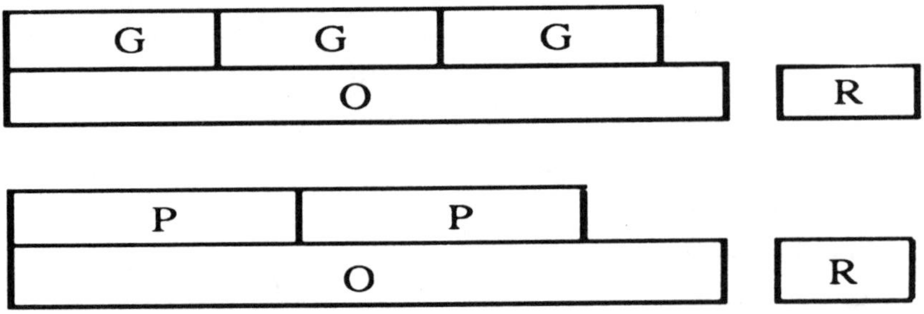

FIGURE 10-34 • *Brian's justification for comparing 3/4 and 2/3*

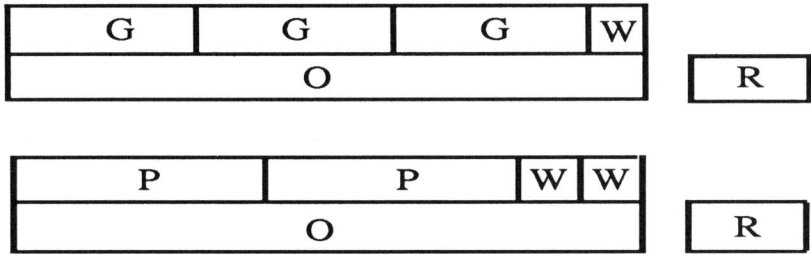

The interviewer then asked Brian to show this with numbers. Brian responded by writing the subtraction problem that is shown in Figure 10-36. Unlike his behavior in the March interview before instruction began, Brian was now able to solve the problem correctly with numbers.

Brian: O.K. There's . . . Um . . . thirds here, fourths . . . I've got to go back down . . . and that's. If I had it the other way [changes the vertical order of the two fractions], you'd subtract and it'd be one-twelfth.

Interviewer: And why would you do it the other way? Why would you write it the other way?

Brian: Cause this [the 3/4] is bigger.

FIGURE 10-35 • *Brian's solution model*

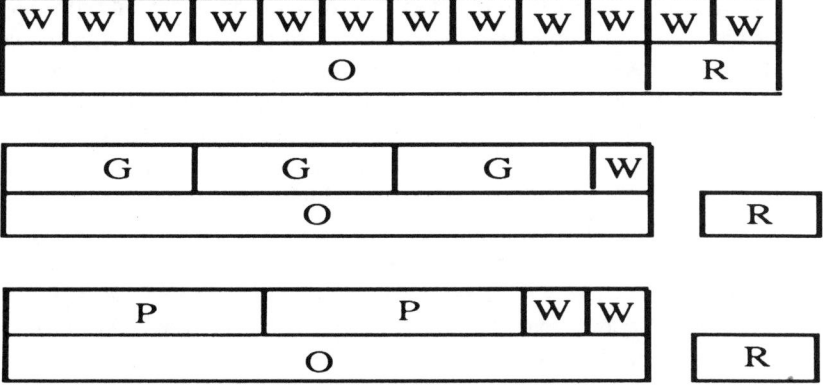

FIGURE 10-36 • *Brian's number solution*

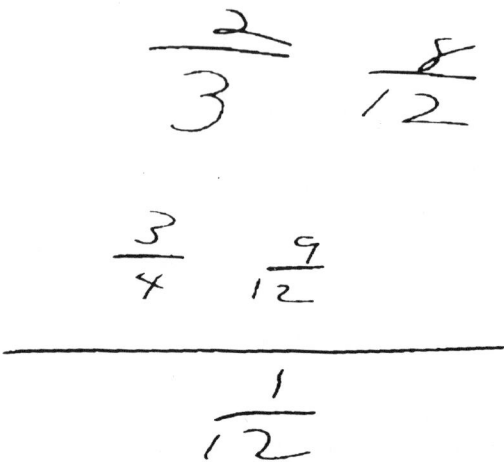

As compared with Brian's earlier solution from the preinterview data, Brian was now able to connect the model he built to his symbolic solution. In both cases, Brian expressed confidence in his answer that the difference was 1/12. Now he expressed confidence in his two approaches, one found by comparing the two models and the other by subtracting the smaller fraction from the larger.

Brian's Classroom Mathematics in Grade 6

Classroom Activity Using Cuisenaire Rods (April 1989)

During the first several lessons of the sixth-grade unit, the children's problem-solving experiences with fractions was focused primarily on representations of fractions. The children were involved in problem activities using Pattern Blocks, Cuisenaire rods, number lines, or geoboards to determine parts of particular units or to construct some mixed number on the basis of a defined unit. The children also completed several activities involving fractions as operators. These activities had provided similar concrete materials for the children to use to solve problems in which they found a particular fractional part of another quantity.

A segment from the transcript of one of these lessons, about midway through the unit, provides an example of Brian's problem-solving activity that might account to some extent for the difference in his numerical representations in the first and the second interviews.

The teacher, Janet, had given each of the children a fraction activity using Cuisenaire rods, which they were to complete, working with a partner. The first problem that Brian and his partner, Ryan, considered is stated as follows:

If we call the train made of "Orange and Red" together 4, figure out the numerical value for red, light green and dark green.

The problem continues, asking for the numerical values of each rod or train that is shorter than the orange and red if the value of orange and red together is 4. The two boys read the problem individually, talked together about their inability to understand it, and then sat quietly, each rereading and thinking alone about possible ways to begin.

Brian: This would be . . .

Brian took an orange and a red rod together to make the train defined as 4, then took out a number of red rods and lined them up to match the train as shown in Figure 10-37.

He then pushed aside the red rods, one at a time, quietly counting the number used.

Ryan: I don't know what I'm doing.

Brian: I know what I'm doing. [This is typical of the confidence that Brian showed in grades 5 and 6.]

Brian took out several light green rods and lined them up to form a train that was equivalent to the red and orange train, as shown in Figure 10-38.

FIGURE 10-37 • *Brian's models of four units*

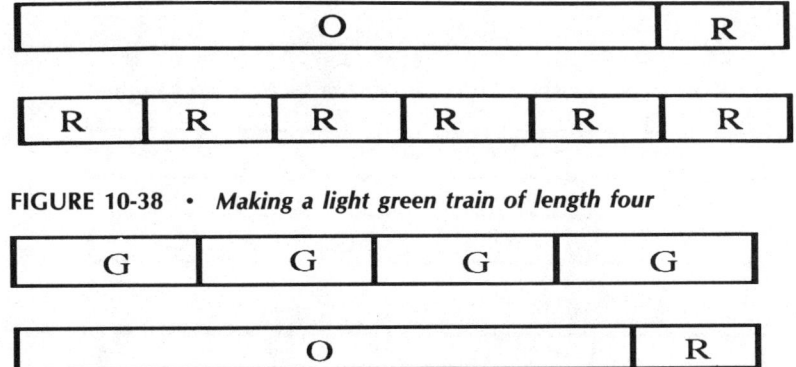

FIGURE 10-38 • *Making a light green train of length four*

Brian: Ryan, I'm trying to figure . . .

Ryan: I don't get it.

Brian then showed Ryan his model of the two trains and said:

Brian: See . . . this thing, the orange and red . . . is 4.

The boys continued to talk quietly [inaudible to the recording]. Then each returned to his individual work. For about three and a half minutes, Brian intently moved about rods in relation to the two trains. During this time, the teacher, Janet, was moving about the room and talking to the other children. The boys worked quietly, so it was easy to hear the teacher talking to another group. Perhaps her comment was also heard by Brian.

Janet: Remember, this is four whole units. This (the train of orange and red) is like a number line from 0 to 4 [Figure 10-39].

Ryan: Oh . . . I . . .

Ryan, eager to attract the teacher's attention, continually raised his hand. Brian, meantime, is trying to determine the proper number name to assign to the red rod, if the orange and red together are being called "four":

Brian: Oh! I know what it is! [He hit his own head with his hand.] It's two-thirds! [That is, the red rod should be given the value 2/3, in order to be consistent with the assignment of the value 4 to the orange and red train.]

Ryan: What's two-thirds? Two-fourths. That's 4 (indicating Figure 10-40) not thirds.

Brian: Pay attention to what I'm doing. Maybe you'll find out [indicating a rather annoyed tone].

FIGURE 10-39 • *Janet's suggestion for visualizing something like this*

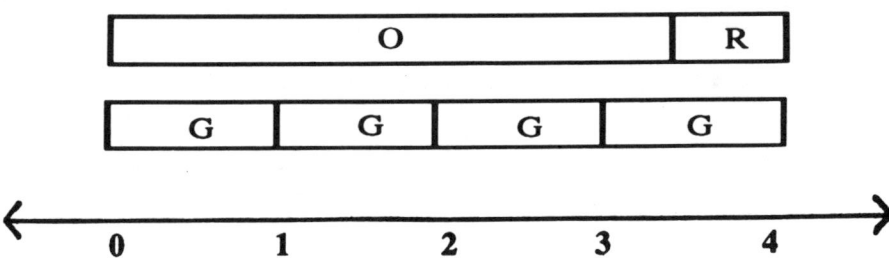

FIGURE 10-40 • *Ryan pointed to this and said, "That's 4."*

O	R

Ryan: Watch . . . [He raised his hand again to attract Janet's attention.]

Brian continued to work with different rods, filling in the values that he assigned to each on his paper. Janet walked over to the boys in response to Ryan's hand, which had continued to be raised.

Ryan: Would this [referring to the red rod shown in Figure 10-41] be two-fourths?

Janet: Think it's two-fourths? Why?

Brian: I think it's two-thirds.

Janet: Two-thirds?

Brian: Because this [referring to his trains of rods indicated in Figure 10-42] equals to 4.

Brian: So I made a number line [points over to his paper]. So if this . . .

Janet: What do you have here? What's that light green [rod]?

Brian: The light green is a whole [indicating Figure 10-43].

Janet: That's 1?

Brian: Yes.

Janet: O.K.

Brian: So this is one, two, out of three [pointing to the red rod in Figure 10-44 as compared to the length of the light green in Figure 10-43].

Janet: Yes . . . O.K.

Brian: So that's what red would equal.

Janet agreed with Brian's reasoning and asked him to explain to Ryan what he had done.

FIGURE 10-41 • *Ryan: "Would this be two-fourths?"*

R

FIGURE 10-42 • *Finding the numerical value for the red rod*

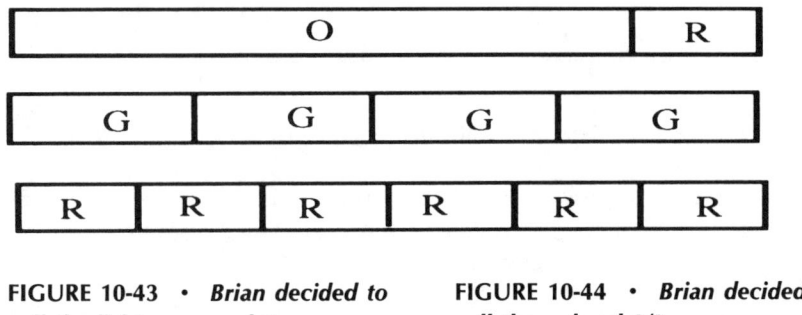

FIGURE 10-43 • *Brian decided to call the light green rod 1*

FIGURE 10-44 • *Brian decided to call the red rod 2/3*

Brian: I said since that's one [points to the light green rod shown in Figure 10-45] . . . two . . . three . . . [referring to the small white rods placed above light green in Figure 10-45) . . . go into it. It splits into three parts and it's two out of three (2/3). [That is to say, Brian has sequentially demonstrated that, if the orange- red train is called "4," then the light green rod must be assigned the value "1," and the white rod must get the value "1/3." All of this is, in fact, correct, if one wishes to preserve additivity of length.]

Janet left the boys to continue the activity for the remainder of the lesson.

In his mathematical activities in grades 5 and 6, Brian's preferred representations were quite concrete, usually either pictures or actual arrangements of wooden blocks. When using those representations, he showed clearly that he understood what he was doing and that he intended to achieve understanding. He would settle for nothing less.

FIGURE 10-45 • *Brian explains why the red rod should be called 2/3*

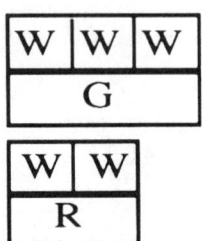

A Two-Year Overview

In our early observations of Brian in grades 5 and 6, we noticed how he built ideas and how he tried to make sense of symbolic representations of the ideas. We also see (in grade 6) an example of classroom problem solving in which Brian is able to figure out how to represent various fractional parts of a unit other than one. The nature of the task and the opportunity to explore possibilities by building models and testing their appropriateness is quite different from those mathematics classroom activities where more of the decisions are made by the teacher, and where, in particular, students are told which materials to use and how to use them. For Brian, the benefit of relatively free exploration can be seen in the June 1989 interview, following classroom instruction (Davis, 1986].

Our observations of Brian during the first two years suggest some patterns in his approach to building solutions to problems. Brian's earlier power came from constructing models, either with pictures or blocks, rather than from symbolically driven solutions to problems. When symbolic solutions began to make sense, the sense-making was derived from the meaningfulness of the solution he constructed.

How might Brian's experiences with mathematics and his approach to the subject be summarized? We offer several observations:

1. Brian likes to figure things out for himself. It might be more accurate to say that, at least in grade 5, *he insisted* on it. In tapes where one sees Brian working with Scott, the difference in their approaches is striking; Scott does whatever the teacher suggests, as in a remark that he makes: "You have to keep the bottom numbers the same." Scott does not ask why, or what this means. Brian always does, and is never satisfied until he can make a concrete model that makes *complete* sense to him.

2. As a general rule, Brian is not effective in taking in the suggestions or remarks of others. He is typically busy building up his own representation of a problem, and unless a suggestion relates directly to the model that Brian is building, it probably will not be heard.

3. Of course, as the tapes show all too clearly, the suggestions that other students — and even the teacher — make to Brian are often incorrect. As some viewers have said, think of the complex task this leaves for Brian: When he hears a suggestion or criticism, it will often be irreconcilable with the representation he is building. He must try to see what meaning, if any, he can find in the suggestion, and if it does not seem to fit in with his representation, he must try to see how he can change his representation so as to make some

match possible, *and he must ask himself which is correct, his model or the other person's suggestion*. This can hardly be an easy task for a 10-year-old.

4. Over the years, Brian's methods do grow in power and sophistication, as when (in one interview) he interprets $1/n$ as necessarily larger than $1/(n + 1)$, because you are sharing with fewer other people.

Concluding Remarks

In all of this, it has literally amazed us to see how much more complexity is involved in "simple" mathematical thinking than we had previously imagined. We are continuing studies of this type, with Brian as well as with other students. We expect to continue to be amazed. As the United States (and other countries) undertake the serious improvement of school mathematics, let no one assume that what is under discussion is some simple matter of responding "4" when you see the stimulus "2 + 2." How children acquire power in thinking about mathematics is a very complicated process.

Examples abound. On other tapes we have, for example, a discussion by four intelligent adults who are trying to find the error in a child's "proof" that 6 = 60. Few viewers of this tape can recognize the essential mistake in the child's conceptualization. Thinking about mathematics is not simple.

There is a second point that demands our attention. The study of Brian—which continues even to the present time—raises a question about typical school programs in mathematics. Brian's experience has been anything but "typical." There have been available to Brian and his classmates many rich opportunities to work with other children on mathematical tasks, to explore mathematics by methods of their own choosing, to seek out meanings in ways that they determine, to talk about methods and strategies, and to build problem solutions cooperatively. In grades 5 and 6 we see Brian and his partner usually using objects or drawing pictures as they worked on the problems. This is not what one typically finds in American schools. For Brian and his classmates there has been an opportunity to deal with mathematics as a *thoughtful* subject. Few children in the United States have such an opportunity. If we want more children to experience mathematics this way, we have our work cut out for us, in no uncertain terms. In particular, we have to help teachers themselves to come to see mathematics as primarily a matter of *thinking, analyzing,* and *planning*.

References

Davis, R. B. (1984). *Learning mathematics: The cognitive science approach to mathematics education*. Norwood, NJ: Ablex.

Davis, R. B. (1986). The convergence of cognitive science and mathematics education. *Journal of Mathematical Behavior, 5*(3), 321-333.

Davis, R. B., & Maher, C. A. (1990). Building representations of children's meanings. In R. B. Davis, C. A. Maher, & N. Noddings (Eds.), *Constructivist views on the teaching and learning of mathematics.* Monograph No. 4, *Journal for Research in Mathematics Education.* Reston, VA: National Council of Teachers of Mathematics.

Landis, J. H. (1990). *Teachers' prediction and identification of children's mathematical behavior: Two case studies.* Unpublished doctoral dissertation, Rutgers University.

Landis, J. H., & Maher, C. A. (1989). Observations of Carrie, a fourth grade student, doing mathematics. *Journal of Mathematical Behavior, 8*(1), 3-12.

Maher, C. A., & Davis, R. B. (1990). Building representations of children's meanings. In R. B. Davis, C. A. Maher, & N. Noddings (Eds.), *Constructivist views on the teaching and learning of mathematics.* Monograph No. 4, *Journal for Research in Mathematics Education.* Reston, VA: National Council of Teachers of Mathematics.

Maher, C. A., Davis, R. B., & Alston, A. (1991). Brian's representation and development of mathematical knowledge: A four-year-study. *Journal of Mathematical Behavior, 10*(2), 163-210.

Maher, C. A., Davis, R. B., & Alston, A. (in press). A teacher's struggle to assess student cognitive growth. In R. Lesh & S. Lamon (Eds.), *Assessing deeper and higher-order understandings of foundation-level mathematical ideas.*

Rosnick, P., & Clement, J. (1980). Learning without understanding: The effect of tutoring strategies on algebra misconceptions. *Journal of Mathematical Behavior, 3*(1), 3-27.

CHAPTER ELEVEN

Mathematics and the Reality of the Student
Bringing the Two Together

ANNA O. GRAEBER

Abstract

Over the last twenty years researchers have documented many ways in which students' mathematical concepts and their beliefs about mathematics (students' mathematics) differ from the concepts and beliefs held by those people society considers more mathematically sophisticated (real mathematics). This chapter gives six examples of the many misconceptions that have been documented as prevalent among students from elementary school through college level. The argument is then made that students' mathematical power would be increased if their understanding of mathematics could be brought closer to real mathematics. Five theories concerning teaching for conceptual change are described: Flavell's steps to equilibrium, Driver's general structure of lesson schemes, Swan and Bell's conflict teaching, the van Hiele phases of learning, and Fischbein's didactical implications of intuition. The instructional strategies and conditions of classroom environment common to the theories are cited. Although there is considerable agreement about what seems to be necessary, and some research to support that theory, barriers exist to the implementation of such instruction. Currently, relatively little information about misconceptions and conceptual change strategies is found in mathematics curriculum materials or in preservice teacher education courses. Current practices in student assessment also inhibit such teaching. Although learners will never be free of misconceptions, we have some ideas about what conditions are likely to help students overcome or at least control some of their misconceptions. The question of how to bring about these conditions remains.

This chapter is based in part on research done in conjunction with grant TEI-8751456 from the National Science Foundation. The opinions and interpretations expressed do not necessarily reflect the position, policy, or endorsement of the National Science Foundation.

It seems somewhat paradoxical that the words *real* and *reality* have so many interpretations, but then the nature of reality has been the subject of much of the world's philosophy. Even within the confines of the mathematics education literature, the concept of *real math* has several different interpretations. For some, real math in the school context is limited to problems that have their source and application in the immediate experience of the student. The cost of five postage stamps at $0.29 per stamp is not an opportunity for real math unless it is the question of a student faced with the need for purchasing five stamps and concerned about the amount of money needed. For others (e.g., Willoughby, Bereiter, Hilton, & Rubinstein, 1981), "making math real for children does not mean simply making it realistic" (p. xvi). It is real if it has meaning and purpose for children. Thus, students engaged in mathematics about fictitious situations or characters may be doing "real math." Booth (1981) and Hart (1981) discuss "child methods"—naive, intuitive, or primitive methods, frequently based on counting or adding on, that children use to solve problems. Some writers would likely claim that these child methods are "real mathematics" in that they are student inventions that are understood by the student. Indeed, Gravemeijer, van der Heuvel, and Streefland (1990) define *realistic* as "not only establishing the connection between reality and the mathematics to be learned, but also creating the possibility for the learners to construct a mathematical reality" (p. vii).

In this chapter, the term *student's mathematics* will be used to describe what students perceive mathematics to be (both its cognitive and its affective dimensions). *Real Mathematics* will be used to describe mathematics that is represented in the shared understanding of "others who are more knowledgeable" (Underhill, 1988, p. 56).

Real Mathematics and Students' Mathematics Differ

A relatively new theme in mathematics education literature deals with the mathematics of students. Indeed, one of the earliest sources that focused on children's mathematical "thought" (as opposed to student errors) was the *Journal of Children's Mathematical Behavior*, begun in the early 1970s. Writers were encouraged to share notions of what "'mathematical thought' means for children, how it develops, and how one might attempt to study it" (Davis & Ginsburg, 1971-192, p. 5).

Since that time a large number of studies have given glimpses of students' mathematical thinking and have provided many examples that illustrate differences between students' mathematics and real mathematics. Although these differences are not omnipresent, they exist at every level from the earliest elementary grades through high school and college and are

Some Student Math in Elementary School

Behr, Erlwanger, and Nichols (1980) studied elementary students' conceptions of the equals sign (=) and found that many elementary students interpreted it as meaning "do something." Shown the sentence $\Delta = 4 + 5$, some 6- and 7-year-olds rewrote it as $\Delta + 4 = 5$, while another asked the interviewer if he read backwards (Behr, Erlwanger, & Nichols, 1980). Many students, reinforced by the countless times the equals sign is used to express the basic addition, subtraction, multiplication, and division facts, continue to rely on this initial interpretation in junior high (Kieran, 1979), high school (Wagner, 1977), and beyond (Clement, 1980). The difficulties this presents in students' ability to make sense of algebraic equations or statements of identity have been documented (e.g., Booth, 1988; Kieran, 1981).

Carpenter, Moser, and Bebout (1988) noted that when students are first presented with the tasks of writing mathematical sentences to represent word problems, "they often see no connection between the informal modeling and counting strategies they use to solve the problems and the number sentences they are taught to write to represent them" (pp. 345–346). Frequently students in elementary school only have opportunity to work with number sentences in the canonical forms of $a + b$ or $a - b$. Carey (1991) notes that these representations do not match the way many children conceive of problems such as "I have seven pennies now, how many more do I need to have twelve pennies?" Most children see the latter as $7 + \Delta = 12$, a number sentence that directly models the implied actions in the word problem. Although mathematicians may see this problem as equally well represented by $12 - \Delta = 7$, young students frequently do not. And some students willing to accept $12 - \Delta = 7$ as a sentence that could be used for the pennies problem do not recognize $54 - 27 = \Delta$ as a representation of a problem for which they will correctly write $27 + \Delta = 54$.

Some Middle School Student Math

Among the most seductive and pervasive student beliefs that go essentially unchallenged until middle school are "multiplication always makes bigger, division always makes smaller." Not until students encounter multiplication and division by fractions and decimals less than one do these beliefs become troublesome. Textbook presentations of algorithms for the operations with fractions and decimals frequently do not reference the underlying meaning of the operation and do not caution students that existing notions about the operations might not hold in the domain of rational numbers (Graeber & Baker, 1991). Many students memorize the procedures, completely or partially, without knowing what they are doing. When computation is treated without meaning, the beliefs frequently become apparent only when the almost universally dreaded word problems arise. Bell, Fischbein, and Greer (1984), Greer and Mangan (1986), and Sowder (1986) have documented students' choice of operations based on the anticipated size of "correct" answers. Greer and Mangan (1986) noted that when the multiplier in a word problem shifted from either a whole number or a decimal greater than one to a decimal less than one, performance fell by about 46% for all groups (from 10-year-olds to teacher trainees) in the study. For some students these beliefs appear to be quite strong (Greer, 1987). Bell, Fischbein, and Greer (1984) applied the phrase "lack of conservation of operation" to students who agree, for example, that to find the number of boxes needed to pack 10 pounds of macaroni into boxes containing 2 pounds each, one divides (10 ÷ 2), but who also argue that to find the number of boxes needed to pack 10 pounds of cookies into boxes containing 0.65 pounds, one multiplies (reasoning that *more* than ten boxes will be needed). Indeed, in one study Sowder (1986, p. 90) found that 30 to 47% of algebra students might be described as lacking conservation of operation.

The differences between students' and real mathematics are not limited to fields involving computation or alphanumeric symbols. Many students believe that of the shapes shown in Figure 11-1, only those labeled *a* and *c* are right triangles. Some students do not recognize *b* and *e* as right triangles because neither has a leg parallel to either the horizontal or vertical (Hershkowitz, 1987). Such students seem to have formed a concept of "right triangle" that includes the condition "one leg horizontal." Similarly, some students have difficulty recognizing isosceles triangles whose base is not horizontal (Hershkowitz, 1987) and even squares with nonhorizontal sides (Usiskin, 1982). The debate continues as to whether these seemingly faulty definitions are constructed from seeing textbook examples in this form (many examples seem to be in stereotypic positions) or whether they reflect a type of "innate sense of horizontal/vertical orientation preference" (see, e.g., Cooper & Krainer, 1990; Hershkowitz, 1987; Zykova, 1969). Still other students claim that because d faces or opens toward the left, it is a "left triangle" and not a right triangle (Pimm, 1987). Such students appear to have applied a common language meaning of "right" to a technical (geometric) domain. Sterotypic images of geometric figures, characteristic perhaps of van Hiele level 0, even seem to dominate students' image of the primitive concept *triangle*. Usiskin (1982) found that one-third of his sample of 2,700 beginning tenth-grade geometry students did not identify "long skinny" triangles (see shape f in Figure 11-1) as triangles.

FIGURE 11-1 • *Shapes for classification as "right triangles" or "triangles"*

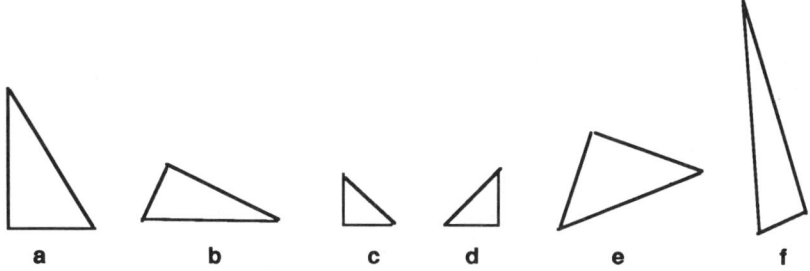

Some Senior High/College Student Math

By now the "six students to one professor problem" must be among the most studied problems in mathematics education. Presented with the direction, "Write an equation for the statement, 'At this university there are six times as many students as professors,' using S for the number of students and P for the number of professors," about 50% of all first-year undergraduate students (Kaput & Sims-Knight, 1983) and 24% of engineering majors (Clement, Lochhead, & Monk, 1981) wrote 6S = P rather than the mathematically correct 6P = S. Researchers have evidence to suggest that these faulty translations come from both a direct word to symbol mapping and from the use of variables as labels (Lochhead & Mestre, 1988). Students using the second mode of thought argue that 6S = P means that for every six students there is one professor. These notions are not far-fetched! The first strategy, a strict left-to-right word-to-symbol mapping, is sometimes taught! The second is believed to be an extension of students' familiarity with statements in which "variables" are treated as labels, e.g., as *3 t* = 1 T, 4 c = 1 pt, 1 km = 1,000 m, etc. (Davis, 1984, pp. 116-117).

One of the most fundamental concepts in calculus is that of a *limit*. A common misconception held by calculus students is that limit is a procedure instead of a number. Students with this misconception identify the limit of a function as the process of the function "approaching a value" instead of the numerical value that is being approached (Heid, 1984). The "limit as approaching" misconception is sometimes coupled with another common misconception about limits—that the value being approached is never attained. Students with this misconception use the language "gets close to but never reaches" to describe the relationship between the process of approaching and the value that is being approached (Tall & Vinner, 1981; Davis & Vinner, 1986; Williams, 1990). Moreover they think that a sequence cannot reach its limit.

Some Student Beliefs about Math

One of the most pervasive views students hold about mathematics is that it is a large set of rules and procedures with no connections. For such students, understanding mathematics means recalling the rules and getting the answer (Skemp, 1978; Underhill, 1988). Such attitudes have been observed among first- and second-grade children (Cobb, 1984), intermediate grade students (Clay & Kolb, 1983; Wheatley, 1984) as well as junior (Frank, 1985) and senior high students (Brown & Cooney, 1985) and teachers (Rector & Ferrini-Mundy, 1986). Data from the Fourth National Assessment of Educational Progress in Mathematics (Lindquist, 1989) indicate that slightly more than 80% of the seventh- and eleventh-graders tested agreed that "There is always a rule to follow in solving mathematical problems" (p. 114). About 50% of those tested agreed that "Learning mathematics is mostly memorizing"; about 20% agreed that "Mathematics is made up of unrelated topics" (p. 114).

Another belief about mathematics shared by many students is that some people just can't do mathematics. Stevenson, Lee, and Stigler (1986) have made us aware that U.S. culture, unlike some others, generally accepts the view that success in mathematics is predominantly a matter of a specific innate ability. The role that effort plays in success in mathematics is felt to be rather minimal. Thus, parents accept a child's poor performance in mathematics, because "I never did well in math either." Recently a student admitted to large university with a mathematics requirement brought suit against the university; he charged that his learning disability made him incapable of passing the required minimal math course and that the university's failure to waive the requirement constituted discrimination. The student's claim was backed by at least one psychologist. Has any such claim ever been lodged against freshman English?

Should Student Math and Real Mathematics Be Brought Together?

To what extent should educators be concerned about the discrepancies between real mathematics and student mathematics? Is there any reason to believe that the conceptual and attitudinal discrepancies that exist between many students' understanding of mathematics and those of mathematicians make a difference?

Most mathematics educators seem to agree that poor mathematics achievement of students in the United States requires changes in the goals and means of mathematics education. They cite data from international studies of achievement in mathematics (McKnight et al., 1987; Lapointe, Mead, & Phillips, 1989), and the Fourth National Assessment of Educational Progress (Lindquist, 1989). However, there are some differences in the directions people suggest for the change.

Some argue that not everyone needs to be a mathematician, but everyone needs some basic mathematical skills. All that is required is that students be able to demonstrate "basic skills." Saxon (1987) notes: "I believe that these unbelievably poor scores are the direct result of trying to teach advanced concepts too early and of neglecting the emphasis of fundamental concepts that can be taught and whose use can be automated by students" (p. 20). He goes on:

> . . . I note a distinct pattern. There has been a search for a panacea, for a magic elixir, for some wonderful way to "teach higher order thought processes" (whatever that means) and a neglect, almost to the point of overt disdain, of trying to automate the fundamental concepts and skills that are necessary for any intelligent application of mathematics.

Effros (1989) argues that "mathematics is above all a language, and its basic elements are first absorbed through drill" (p. 36).

Others argue strongly that weaknesses in algorithmic skills need no longer prevent students from understanding concepts. Many commissions and reports (e.g., American Association for the Advancement of Science, 1989; Mathematical Sciences Education Board, 1990) have argued that the age of computer and calculator technology requires new skills, new topics, and new emphases. These reports also claim that in the near future women and minorities—groups who as students typically have not done well in mathematics—will make up a larger percentage of the work force than at present. Because more jobs will require higher level mathematical skills, it behooves the nation to make efforts to see that these populations take and become successful with more mathematics. Problem solving, reasoning, valuing mathematics and becoming confident in one's ability to do mathematics are primary goals. Thus the *Curriculum and Evaluation Standards for School Mathematics* (National Council of Teachers of Mathematics, 1989) call for teaching that is developed from problem situations.

The admonitions to teach from the problem situation are based in part on currently favored learning theories. These theories are compatible with the notion that if real mathematics and student mathematics are brought together, both more favorable attitudes toward mathematics and increased achievement in mathematics will follow.

For example, constructivists argue that students do attempt to bring order and meaning to their experiences. Carpenter, Hiebert, and Moser (1983) have studied preschool and kindergarteners' problem-solving skills and have suggested that the understanding and sense-making that students bring to school may be undermined by a school curriculum that attempts to force a match between students' understandings and standard symbolic representations. If students see their early experiences with symbols as nonsensical, it is no wonder that they begin to see mathematics as a senseless bunch of rules that must be learned. Further, since mathematics does not fit with their previous sense, doing mathematics must be synonymous with understanding mathematics.

The notion that students must be allowed to link meaning with the symbols used in real mathematics is also argued forcibly by Hiebert and LeFevere (1986) and Wearne and Hiebert (1988). They note that when instruction proceeds without allowing students to make such links, the only recourse for many students is to abandon reason and to memorize. This results in the commitment to memory of many rules and algorithmic procedures that may easily become confused. Such memorization also leaves students without a means of checking the sensibility of their recalled rules and routines; in a sense such students are conceptually stranded. They have given up on the idea that their innate sensibility (or intuition) is helpful, and often they have not experienced any way of connecting back to it. When their rules lead to errors, whether because the wrong rule is used or because a rule is remembered incorrectly, their failure can result in a further sense of powerlessness and poor self-concept. The notion of "learned helplessness" in mathematics has been described by Gentile and Monoco (1986).

Could the bringing together of real mathematics and student mathematics help to avoid this learned-helplessness trap? Among the benefits to be accrued by bringing together real mathematics and student mathematics is the notion that in real mathematics one can "help oneself" as opposed to yielding to reliance on outside authority.[1] If students hold meanings for the symbols used in mathematics, they can perform checks on the reasonableness of their recalled procedures without resorting to the teacher, the text, a student viewed as "superior," or other outside authority. If students are permitted to learn mathematics in a manner that fosters their sense-making, then the "mathematical power" the *Curriculum and Evaluation Standards for School Mathematics* (National Council of Teachers of Mathematics, 1989) aim for may be achievable.

Other educators with a cognitivist or an information-processing view of learning have noted that students trying to reconcile notions that seem inconsistent to them may end up compartmentalizing their knowledge (Posner, Strike, Hewson, & Gertzog, 1982). Indeed, studies of students' thinking have noted areas in which students hold contradictory mathematical notions. For example, Wilson (1990) describes inconsistencies in students' definitions of basic geometric figures and the drawings they select as examples of those figures. Similarly, Tall (1990) identified inconsistencies in university students' definitions of the limit of a sequence and their applications of the definition to specific examples, such as 0.9 + 0.09 + .009 + ... Thus inconsistencies, whether consciously perceived or not, may lead students to compartmentalize concepts and procedures. The notion of compartmentalized knowledge may explain, for example, why preservice teachers indicate that multiplication "does not always make bigger" but act as if it did when solving word problems. Their explicit knowledge of rules is not connected with the knowledge they apply in solving problems.

Is there a need to be concerned about compartmentalized knowledge? Greeno (1978) has suggested three criteria for evaluating degree of understanding: coherence (internal integration), connectedness (to other concepts), and correspondence. Researchers (e.g., Larkin, 1977) noted that experts recall information in "chunks" or "bursts," whereas novices' recall of information was more randomly distributed over time. If students are to be facile at problem solving, facility with recall of pertinent information seems crucial. If the goal of instruction is understanding and problem-solving ability, there appears to be an argument for the superiority of coherent and connected knowledge—qualities of knowledge more typical of real mathematics than of the knowledge many students possess.

What Changes Are Needed If Real Mathematics and Student Mathematics Are to Be Brought Together?

This chapter, like many of the existing calls for change in mathematics education, claims that changes in pedagogy and curriculum, in assessment, and in societal expectations are all needed if we are to bring about any reconciliation of student mathematics and real mathematics.

In this chapter the necessary changes are approached from the specific viewpoint of teaching that evokes, recognizes, and builds upon students' perceptions (some might identify this with teaching for understanding). Those who see learning as necessitating conceptual growth, perhaps as well as behavioral change, view all learning as involving conceptual change. Not only must students' existing misconceptions change, but in areas where they have no strong conceptions new ideas must be fitted into the existing relationships in the students' conceptual scheme.

Changes in Pedagogy and Curriculum

Although it might be nice to think that if we could just get teaching at the elementary level "straightened out," all misconceptions would be prevented, this is a naive point of view. Misconceptions such as "equals as becoming" or "multiplication always makes bigger" are not primarily the result of direct instruction. Rather, they are more or less "natural" conclusions based on early experience. As Fischbein, Derri, Nello, and Marino (1985) have argued, it is unlikely that we will begin the teaching of multiplication in the set of rational numbers, and early whole number experience will give students the idea that "multiplication makes bigger." Thus there is a curricular dilemma, in that the order in which it is logical to teach this topic increases the chances of the formation of the misconception. Some even argue that the "frame of reference" misconception is not merely the result of experience with figures that have their base parallel to a side of the page or screen, but that there is some innate (related to the body axis) or culturally ingrained preference for figures in a horizontal/vertical orientation (Cooper & Krainer, 1990; Fisher, 1983; Young, 1982). Further, new misconceptions are likely to arise as new topics, new approaches, and new technologies are used. For example, Goldenberg (1987) has reported on misconceptions that appear to be fostered in computer graphing environments.

Thus it appears likely that teachers will always be in the position of changing students' ideas. Indeed, some have argued (Bell, personal communication; Behr & Harel, 1990) that all learning involves some conflict and change. What do theory and research suggest for the teacher faced with the need to have an impact on students' misconceptions? This section considers some of the frameworks suggested for those designing learning experiences aimed at changing students' ideas. Some (Driver 1987; Fischbein, 1987; Swan, 1983) have been developed precisely for the purpose of having an impact on misconceptions. Others, such as that of Flavell (1977) and the van Hieles (Fuys, Geddes, & Tischler, 1988) were derived to answer the more generic question, "What is needed for cognitive growth?"

Although the developers of several of the models included in the chart cite the work of Piaget and Flavell as a starting point for their thinking, others, such as Driver (1987) and Van Hiele (Fuys, Geddes, & Tischler, 1988), claim a more eclectic or pragmatic approach to building their framework. What will probably be of most interest are the common features shared by the models. The remainder of this section includes comments on each model or framework and then a summary of the suggestions that seem to be common across the models.

Approaches to Cognitive Conflict/Cognitive Growth

The notion of "cognitive conflict" is generally attributed to Inhelder, Sinclair, and Bovet (1974),[2] who investigated children's progress in attaining concepts such as conservation of matter, quantity, length, and number.

The training methods they developed to encourage the development of these concepts were summarized as

> constructed to encourage the child's activity and to elicit the coordination and differentiation of thought patterns that are characteristic of the different levels of development. No attempt was made to lead the child through a series of preprogrammed steps toward the correct solution of a problem. The procedures provided the student with a series of situations which favored their apprehension of the experimental facts and which led to numerous comparisons and conflicts between the subjects' predication and ideas and the actual outcome of certain manipulations. (Inhelder, Sinclair, & Bovet, 1974, p. 243)

The authors commented that such training situations led to "temporary disequilibria, which provide the impetus for new constructions, . . . , the disequilibria are experienced by a child as conflicts or contradictions" (pp. 258–259). The authors traced four steps in the development of conservation. In the first step, children keep two different modes of reasoning completely separate. For example, they will agree that a one-to-one match of rods implies that two "roads" made by the chains of rods are the same length. But if one of the "roads" is then configured as a zig-zag (see Figure 11-2), children will judge length on the basis of, say, the horizontal distance traversed, and deny that two roads with the same number of rods are the same length. The second phase is marked by children's apparent awareness of and struggle to understand the discrepancy between the two modes (see Figure 11-2).

The third phase included some attempt to arrive at a compromise between the two situations, although these compromises are incomplete or partial. A final stage is reached when the child is capable of complete understanding. The authors conclude that experience, in particular experience of discrepancies between one's predictions and actual outcomes, is therefore an important factor in the acquisition of knowledge.

The phases through which children presumably move if they are to progress cognitively has perhaps been best summarized by Flavell (1977). He describes these as (1) noticing both of the apparently conflicting elements in a situation; (2) interpreting and appreciating the two as conflicting;

FIGURE 11-2 • *Rod configuration in conservation of length task*

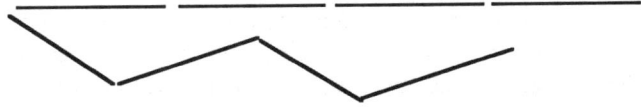

(3) attempting an explanation for the difference rather than "clinging defensively to" the "initial belief or refusing to have anything more to do with the problem" (p. 242); and (4) constructing a new conceptualization that accommodates both of the elements, thereby reaching equilibrium.

The cognitive development theory espoused by Flavell, Inhelder, Bovet, and Sinclair is an attempt to describe the processes involved in cognitive change. They say little directly about instruction designed to promote or enhance such change. The perspective suggests to some that the pedagogy needed to help students overcome misconceptions would enable students to evoke two conflicting ideas, to become cognizant of the ideas as conflicting, to attempt to resolve the conflict, and then to establish a new view. Although the proponents of "conflict teaching" do not always trace their roots to the work of Flavell, some of the lessons developed follow the general outline suggested by him.[3] This can be seen in the examples given in the sections below on Driver's theory and "conflict teaching."

Driver's General Structure of Lesson Schemes Rosalind Driver and associates at the Centre for Studies in Science and Mathematics Education at the University of Leeds, England, have worked on curriculum development primarily in the area of science education. This group approaches its work from a constructivist point of view, noting that curriculum "is not that which is to be learned, but a programme of learning tasks, materials and resources which enable students to reconstruct their models of the world to be closer to those of school science" (Driver, 1989, p. 82). Driver argues that teachers must have a coherent grasp of the subject if there is to be any chance for content to be transmitted in an effective way.

The development of units is guided by consideration of (1) what should be taught, (2) students' naive ideas about the content, (3) the constructivist notions that conceptual change occurs as the result of active processing of information by the learner, and (4) teachers' experiential knowledge of what works and what does not work in the classroom. The units reflect an overall sequence of five phases:

- *Orientation*, to establish purpose and motivation, catch attention
- *Elicitation*, to make explicit students' current ideas
- *Restructuring*, to exchange views, develop and resolve conflicts, construct new conceptions, and evaluate new conceptions
- *Application*, to use new conceptions in a variety of situations
- *Review*, to reflect on the change in thinking that has taken place

Although many writings of Driver do not refer directly to the work of Flavell or of Inhelder, Sinclair, and Bovet, the influence of cognitive science seems clear.[4] The phases in Driver's units do correspond fairly well with

Flavell's steps to equilibrium (see Figure 11-3). Some of the embellishments in Driver's phases are clearly motivated by the application to learning. For example the "Apply" phase, though not an explicit part of Flavell's steps to equilibrium, does seem particularly relevant to teachers who are trying to show applications, provide the student with proficiency in using the concept, and teach for transfer (see Figure 11-3).

Driver has also enumerated strategies that might be used in the restructuring phase. The intent of the restructuring phase is to help students clarify and exchange their existing ideas, expose them to a conflict situation, help them construct new ideas, and then evaluate the usefulness of those new ideas. Driver (1987) criticizes approaches that rely heavily on the emergence of counterexamples alone to induce change. She argues that restructuring of a new conception is an important aspect of conceptual change. Some of the strategies that Driver suggests might be used to accomplish this end are listed in Figure 11-3. These strategies have been defined elsewhere (Driver, 1987) and some applications to mathematics education have been suggested (Graeber & Johnson, 1990).

Driver has also listed the classroom conditions that her experience and reading of the literature suggest facilitate an effective conceptual change classroom. These include the provision of a nonthreatening environment where students feel comfortable in expressing and sharing their views. In such an environment, teachers and other students must suspend their evaluation of students' initial concepts. Attention to techniques that help students think about their thinking—that is, attention to metacognition—is seen as necessary. Thus the lesson sequence includes a "review" component, but it is not the usual notion of review found in the United States, the idea of merely "coming back to." Rather, it involves the students in looking back at their own changes in thinking. The students revisit the concepts they held in the elicitation phase of the lesson and compare and contrast them with their present conceptions. The creation of posters of students' conceptions, the keeping of journals, or the production of pre- and postinstructional concept maps are means of promoting this reflection on thinking. Driver also believes that talk is important in helping students formulate ideas. Thus she feels that small-group work is an important component of a conceptual-change-facilitating classroom. The context of learning tasks should also be meaningful, surprising, unique, or interesting to assist in capturing student interest, stem the notion that science is divorced from life outside school, and help students to see applications of concepts.

Conflict Teaching The concept of conflict teaching is one that has been espoused, tested, and advocated by Malcolm Swan, Alan Bell, and a number of Bell's students. As defined by Swan (1983), conflict teaching includes four phases that closely parallel Flavell's steps to equilibrium: intuitive phase,

FIGURE 11-3 • *Ideas about changing ideas*

Flavell's (1977) steps to equilibrium	Driver's (1987) general structure of lesson schemes	Swan's (1983) "Conflict lesson sequence" (Bell, Brekke, & Swan, 1987)	Van Hiele phases of learning (Fuys, Geddes, & Tischler, 1988)	Fischbein's (1987) didactical implications re intuition
Notice both of the apparently conflicting elements.	Orient	Intuitive phase	*Inquiry*: Student and teacher engage in conversation to determine the student's prior knowledge and to orient the student to the topic.	Student must accept that while absolutely certain, he was wrong.
Interpret and appreciate these differences as conflicting.	Elicit	Conflict phase	*Directed orientation*: Activities that reveal structures to the student.	Don't let student cope on her own. Make clear intuition is natural: many have these difficulties.
Attempt an explanation.	Restructure Extend Differentiate Experiential bridging Import new model Unpack Progressive shaping Construct alternative	Resolution phase →	*Explication*: Students express and exchange views. Concepts	Analyze the source, realize the more general, formal structure. Start as early as possible preparing student for the formal meaning.
Come up with a better conceptualization.	Apply	Reinforcement phase	*Free orientation*: Concept applied in complex tasks.	Naive interpretations may never be eradicated; in the long run a secondary intuition may develop.
	Review	(Retrospection)	*Integration*: Students review and summarize learnings.	

226

conflict phase, resolution phase, and reinforcement phase. In conflict teaching the intuitive phase involves activities that lead students to expose and perhaps even apply their intuitive ideas. The *conflict phase* is designed to introduce alternative conceptions, either from students directly or by applying some new methodology or interpretation that leads to doubt and other conclusions. In the resolution phase, students come to recognize the inconsistencies exposed earlier and to identify the "better" conception and its usefulness. In early writings Swan, Bell, and others identified the fourth and a final phase as a reinforcement phase in which the new concept is used. More recently Bell, Brekke, and Swan (1987) have included retrospection ("review" in the sense used by Driver) in this phase. Indeed, Bell, Brekke, and Swan claim that review of the change in ideas is a very powerful step without which full benefit of conflict teaching does not take place.

Bell and his students have conducted numerous studies comparing conflict teaching with more traditional direct (telling) methods. In most of these studies the mathematical topics involved (multiplication with positive rationals less than one, operations with signed numbers) are prone to common misconceptions. In almost every case the conflict method was more effective than the traditional method, although the results (in terms of student mastery of the concept) were never as spectacular as Bell or many mathematics educators would like them to be. Whether this is a reflection of the persistence of misconceptions or a true weakness in the methodology is unclear.

The Van Hiele Phases of Learning Many educators are familiar with the van Hiele stages of learning in geometry. Fewer may be aware that the van Hieles also proposed phases of learning designed to help students move from one level of understanding to the next (Fuys, Geddes, & Tischler, 1988). The five phases of learning proposed were as follows:

1. *Inquiry/information*: Teacher and students engage in conversation to elicit pupils current conceptions and to set the stage.
2. *Directed orientation*: Students explore through teacher-designed activity. Activities should reveal the concepts appropriate to the van Hiele level for which students are reaching.
3. *Explication*: Students express observations and exchange ideas. Teacher helps by giving appropriate vocabulary.
4. *Free orientation*: Students engage in more complex application tasks.
5. *Integration*: Students review and summarize what they have learned. It is suggested that instruction involving this sequence of phases will help students progress from one level of thought to the next.

Fischbein's Didactical Implications Fischbein, in his book *Intuition in Science and Mathematics* (1987), discusses sources of intuition and emphasizes intuition as both a valued and a necessary aspect of endeavors in mathematics (as well as science) and as something that is prone to lead one astray. Although Fischbein describes numerous faulty intuitions in mathematics, his book gives more insights into his concerns about teachers' need to deal with the affective aspects of dethroning misconceptions than with the strategies for affecting the conceptual change. Fischbein reasons that we do not want students to lose faith in their own sense-making or to say that mathematics is unreasonable. Yet, he notes, when students face conflict between their intuition and other persuasive evidence, a change in their concept requires them to accept that "while being absolutely convinced about the truth of a certain idea," they were "in fact wrong" (Fischbein, 1987, p. 37). Unless students are aware of the fact that their intuitions are common and reasonable, they may give up trying to use their common sense or intuition. Thus Fischbein argues that it is not desirable to "let the students cope by" their "own means with these intuitive difficulties" (p. 38). He urges teachers to make clear to students that everyone, including great mathematicians, encounters difficulties between intuitive beliefs and logically derived results. Teachers must encourage students both to value their intuition (to initiate solutions and check results) and to question it (is intuition leading us astray?).

Fischbein does not "consider that naive intuitive interpretations may be eradicated altogether. The problem is to help the student understand the logical structure of the new conceptions and their superiority. . . . It is possible that, in the long run, the new logically based interpretation will generate a new intuitive acceptance—a secondary intuition" (p. 192). Thus Fischbein warns that even the best of teaching strategies may not eliminate misconceptions, and he stresses the need for students to be aware of their intuitions and to monitor their work on problems so as to check the influence of these misconceptions.

Derived Teaching Strategies

A review of the teaching strategies outlined here suggests that teaching for conceptual change probably requires attention both to the content-oriented pedagogical strategies and to classroom climate. All of the strategies outlined emphasize beginning by eliciting the students' present ideas, provoking doubt about or conflict among these ideas, encouraging the restructure or construction of new ideas, providing opportunities to practice with (apply) the new concept, and stimulating the students' reflection on their own history of the concept.

The authors cited also describe the type of context in which it likely that one can expose one's ideas, subject them to scrutiny by others, change them, study the evolution of ideas, and still remain confident in one's common sense. These attributes include a nonthreatening environment that

- Does not make premature assessments of a right/wrong nature, but permits pupils to share ideas
- Makes clear the commonsense nature of what we here have called misconceptions
- Reinforces the contributions of intuition while at the same time stressing its limitations
- Encourages students to monitor their own thinking and their changes in thinking

Implications for Instructional Materials

With a few exceptions (e.g., Freeman & Porter, 1989), mathematics educators have noted that curriculum materials, especially textbooks, have considerable influence on what and how teachers present in the classroom (see, e.g., Dossey, Mullis, Lindquist, & Chambers, 1988). Curriculum materials could presumably aid teachers and students if they facilitated connections between symbols and their meaning, and made explicit the ideas students are likely to bring to a subject and the reasons for these ideas. Textbooks could also suggest experiences for testing these ideas in a new domain or noting conflicts between these ideas and the concept being explored. If readers believe that current textbooks contain such information, they should select a misconception from the list above and trace the concept through a set of textbooks. They may find difficulties noted, but typically teachers are left on their own to find teaching strategies for these concepts or simply advised that much practice is needed. Rarely will they find explanations for misconceptions, let alone a rationale for their existence, or a note to students indicating the reasonableness of a common misbelief.

Some Needed Changes in Assessment

The literature on misconceptions sends two clear messages related to assessment of students' understanding. *First, purely computational tasks frequently do not uncover misconceptions.* This is true because such tasks are frequently accomplished rotely without any attempt on the student's part to bring reason to bear on them. Thus, in a computational setting, students readily record $0.5 \times 0.5 = 0.25$. *Second, even students' explicit knowledge can hide misconceptions.* Students and preservice teachers asked whether multiplication always makes bigger may say "no," but when they attempt to solve a word problem with a decimal divisor less than one, they frequently multiply (Tirosh & Graeber, 1989). Students may state a definition of a rectangle that includes the notion of square but fail to select squares as instances of rectangle (Wilson, 1990).

Thus it seems that any assessment used in a setting that values the bringing together of "real mathematics" and students' mathematics must go beyond the one or two dimensional type snapshots many current assessment schemes employ. Not only do these single faceted assessments allow misconceptions to go undetected, but as is frequently argued, they send messages to both students and teachers about the type of performance that is valued. If understanding and application are to be valued, assessment must include tasks that require application of concepts and skills. Assessment of thinking, as described in Chapter 12 of this volume by Ginsburg, Jacobs, and Lopez, and tasks that require application [see, e.g., Chapter 13 of this volume, by de Lange) are required.

Changes in Societal Expectations

As the current attempts at curriculum reform have acknowledged, changes in schools are dependent upon attitudes and expectations in the society at large. Until parents and newspaper editors understand that mathematics is more than basic facts and standard algorithms for common arithmetic and algebraic operations, I fear that the prospects for teaching for conceptual change are poor.

Giving up our almost exclusive reliance on easy to administer multiple choice tests and requiring teachers to gain sufficient familiarity with not only the content of mathematics but the "pedagogical content knowledge" (Shulman, 986) is contrary to current trends we see in public policy. Although there are efforts in California and Maryland to introduce performance based assessment, teachers' and public acceptance of these tests are yet to be won. State legislatures and boards that have severely restricted or virtually eliminated "methods courses" from credits counting toward degrees make it difficult to include in preservice teacher preparation the very aspects of pedagogy ("knowledge of the content, the most useful forms of representation, . . . powerful analogies or an understanding of what makes the learning of specific topics easy or difficult . . . , the conceptions and preconceptions that students . . . bring with them"; Shulman, 1986, p. 9) that seem necessary to teachers who want to bring together the mathematics of the student and real mathematics.

Recent studies illustrate that children's mathematics often differs from real mathematics. While perfect alignment is unlikely (even in the long run of reform or over years of education for a given student) a look at most current curricula and modes of instruction suggest that much can be done to bring the two somewhat closer together. While such attempts are not seen as a panacea or an elixir, currently accepted views of learning imply, and bits and pieces of research hint, that bringing the two together can result in improved attitudes toward and achievement in mathematics. The question remains — can we foster the conditions likely to bring this about?

Notes

1. *Editors' note*: A striking demonstration of this difference is presented in Carolyn A. Maher and Alice Alston "Is Meaning Connected to Symbols?" Interview with Ling Chen, *Journal of Mathematical Behavior*, *8*(3), 1989, pp. 241-248.

2. *Editors' note*: Of course, it built upon earlier work by Leon Festinger and Alex Bavelas, and all work on changing conceptualizations owes a large debt to Piaget.

3. *Editors' note*: Indeed, much of the pedagogical work in this spirit actually *precedes* Flavell. In fact, promoting student cognitive growth by identifying and confronting student misconceptions (or limited conceptions) was first developed, not by psychologists, but by practicing teachers. See, e.g., the discussion on "torpedoing" in Robert B. Davis, "Discovery in the Teaching of Mathematics," in Lee S. Shulman & Evan R. Keislar (Eds.), *Learning by Discovery: A Critical Appraisal* (Chicago: Rand McNally, 1966), pp. 117-119).

4. *Editors' note*: Driver's work builds heavily on some unpublished work of Jack Easley, with whom she studied.

References

American Association for the Advancement of Science. (1989). *Science for all Americans*. Washington, DC: Author.

Behr, M., Erlwanger, S., & Nichols, E. (1980). *How children view equality sentences*. PMDC Technical Report No. 3. Tallahassee: Florida State University. (ERIC Document Reproduction Service No. ED 144 802).

Behr, M., & Harel, G. (1990). Students' errors, misconceptions, and cognitive conflict in application of procedures. *Focus on Learning Problems in Mathematics*, *12* (3, 4), 75-84.

Bell, A., Brekke, G., & Swan, M. (1987). Misconceptions, conflict and discussion in the teaching of graphical interpretation. In J. Novak (Ed.), *Proceedings of the Second International Seminar on Misconceptions and Educational Strategies in Science and Mathematics* (Vol. 1, pp. 45-58). Ithaca, NY: Cornell University.

Bell, A., Fischbein, E., & Greer, B. (1984). Choice of operation in verbal arithmetic problems: The effects of number size, problem structure and content. *Educational Studies in Mathematics*, *15*, 129-147.

Booth, L. (1981). Child-methods in secondary mathematics. *Educational Studies in Mathematics*, *12*, 29-41.

Booth, L. (1988). Children's difficulties in beginning algebra. In A. Coxford (Ed.), *The ideas of algebra, K-12* (1988 Yearbook of the NCTM, pp. 209-232). Reston, VA: National Council of Teachers of Mathematics.

Brown, K., & Cooney, T. (1985). The perspectives of students. In S. K. Damarin & M. Shelton (Eds.), *Proceedings of the Seventh Annual Meeting of the North American Chapter of the International Group for the Psychology of Mathematics Education* (pp. 34-37). Columbus: Ohio State University.

Carey, D. (1991). Number sentences: Linking addition and subtraction word problems and symbols. *Journal for Research in Mathematics Education, 22*, 266-280.

Carpenter, T. P., Hiebert, J., & Moser, J. M. (1983). The effect of instruction on children's solutions of addition and subtraction word problems. *Educational Studies in Mathematics, 14*, 55-72.

Carpenter, T. P., Moser, J. M., & Bebout, H. C. (1988). Representation of addition and subtraction word problems. *Journal for Research in Mathematics Education, 19*(4), 345-357.

Clay, G. V., & Kolb, J. R. (1983). The acquisition of meanings for arithmetic symbols. In J. C. Bergeron & N. Herscovics (Eds.), *Proceedings of the Fifth Annual Meeting of the North American Chapter of the International Group for the Psychology of Mathematics Education* (Vol. 1, pp. 243-249). Montreal: University of Montreal.

Clement, J. (April, 1980). *Algebra word problem solutions: Analysis of a common misconception* (based in part on a presentation given at the annual meeting of the American Education Research Association, Boston). Amherst, MA: Hasbrouck Laboratory, Cognitive Processes Research Group.

Clement, J., Lochhead, J., & Monk, G. (1981). Translation difficulties in learning mathematics. *American Mathematical Monthly, 88*, 286-290.

Cobb, P. (1984). The importance of beliefs and expectations in the problem solving performance of second grade pupils. In J. Moser (Ed.), *Proceedings of the sixth annual meeting of the North American Chapter of the International Group for the Psychology of Mathematics Education* (pp. 135-140). Madison: University of Wisconsin.

Cooper, M., & Krainer, K. (1990). Children's recognition of right-angled triangles in unlearned positions. In G. Booker, P. Cobb, & T. Mendicuti (Eds.), *Proceedings of the Fourteenth International Conference for the Psychology of Mathematics Education* (Vol. 2, pp. 227-234). Mexico.

Davis, R. B. (1984). *Learning mathematics: The cognitive science approach to mathematics education*. London: Routledge.

Davis, R., & Ginsburg, H. (1971-1972). Explanation of this journal. *Journal of Children's Mathematical Behavior, 1*(1), 5.

Davis, R. B., & Vinner, S. (1986). The notion of limit: Some seemingly unavoidable misconception stages. *Journal of Mathematical Behavior, 5*(3), 281-303.

Dossey, J., Mullis, I., Lindquist, M., & Chambers, D. (1988). *The mathematics report card: Are we measuring up?* (Report No. 17-M-01). Princeton, NJ: Educational Testing Service.

Driver, R. (1987, July). Promoting conceptual change in classroom settings: The experiences of the children's learning science project. In J. Novak (Ed.), *Proceedings of the Second International Seminar on Misconceptions and Educational Strategies in Science and Mathematics Education* (Vol. 2, pp. 97-107). Ithaca, NY: Cornell University.

Driver, R. (1989). Changing conceptions. In P. Adey, J. Bliss, J. Head, & M. Shayer (Eds.), *Adolescent development and school science* (pp. 79-99). Philadelphia: Falmer Press, Taylor & Francis.

Effros, E. (1989, April 12). When "Everybody Counts," perhaps no one will. *Education Week*, *8*, 36.
Fischbein, E. (1987). *Intuition in science and mathematics: An educational approach*. Boston: Kluwer Academic.
Fischbein, E., Derri, M., Nello, M. S., & Marino, M. S. (1985). The role of implicit models in solving problems in multiplication and division. *Journal for Research in Mathematics Education*, *16*(1), 3-17.
Fisher, N. (1983). Some perceptual influences in learning geometry. In R. Hershkowitz (Ed.), *Proceedings of the Seventh International Conference for the Psychology of Mathematics Education* (pp. 218-222). Rehovat, Israel: Weizman Institute of Science.
Flavell, J. (1977). *Cognitive development*. Englewood Cliffs, NJ: Prentice-Hall.
Frank, M. L. (1985). A framework for problem solving. In S. K. Damarin & M. Shelton (Eds.), *Proceedings of the Seventh Annual Meeting of the North American Chapter of the International Group for the Psychology of Mathematics Education* (pp. 80-85). Columbus: Ohio State University.
Freeman, D., & Porter, A. (1989). Do textbooks dictate the content of mathematics instruction in elementary schools? *American Educational Research Journal*, *26*, 403-421.
Fuys, D., Geddes, D., & Tischler, R. (1988). The van Hiele model of thinking in geometry among adolescents. *Journal for Research in Mathematics Education Monograph*, *3*. Reston, VA: National Council of Teachers of Mathematics.
Gentile, J. R., & Monoco, N. M. (1986). Learned helplessness in mathematics: What educators should know. *Journal of Mathematical Behavior*, *5*, 159-178.
Goldenberg, E. P. (1987). Believing is seeing: How preconceptions influence the perception of graphs. In J. Bergeron, C. Kieran, & N. Herscovics (Eds.), *Proceedings of the Eleventh International Conference for the Psychology of Mathematics Education* (Vol. 1, pp. 282-288). Montreal, Canada.
Graeber, A. O., & Baker, K. (1991). Curriculum materials and misconceptions concerning multiplication and division. *Focus on Learning Problems in Mathematics*, *13*, 25-38.
Graeber, A. O., & Johnson, M. (1990). *Insights into secondary students' understanding of mathematics*. Unpublished manuscript, University of Maryland at College Park, Center for Mathematics Education.
Gravemeijer, K., van der Heuvel, M., & Streefland, L. (1990). *Context-free productions, tests, and geometry in realistic mathematics education*. Utrecht, Netherlands: State University of Utrecht, Research Group for Mathematical Education and Educational Computer Centre.
Greeno, J. (1978). Understanding and procedural knowledge in mathematics instruction. *Educational Psychologist*, *12*(3), 262-283.
Greer, B. (1987). Nonconservation of multiplication and division involving decimals. *Journal for Research in Mathematics Education*, *18*, 37-45.
Greer, B., & Mangan, C. (1986). Choice of operation: From 10-year-olds to student teachers. *Proceedings of the Tenth International Conference for the Psychology of Mathematics Education* (pp. 25-30). London: University of London Institute of Education.
Hart, K. (1981). *Children's understanding of mathematics: 11-16*. London: Murray.
Heid, K. (1984). An exploratory study to examine the effects of resequencing skills and concepts in an applied calculus curriculum through the use of the microcomputers. *Dissertation Abstracts International*, *46* (University Microfilms #85-12, 190).
Hershkowitz, R. (1987). The acquisition of concepts and misconceptions in basic

geometry—or when "a little learning is a dangerous thing." In J. Novak, (Ed.), *Proceedings of the Second International Seminar on Misconceptions and Educational Strategies in Science and Mathematics* (Vol. 3, pp. 238-251). Ithaca, NY: Cornell University.

Hiebert, J., & LeFevere, P. (1986). Conceptual and procedural knowledge in mathematics: An introductory analysis. In J. Hiebert (Ed.), *Conceptual and procedural knowledge: The case in mathematics*. Hillsdale, NJ: Lawrence Erlbaum.

Inhelder, B., Sinclair, H., & Bovet, M. (1974). *Learning and the development of cognition*. Cambridge, MA: Harvard University Press.

Kaput, J., & Sims-Knight, J. (1983). Errors in translations to algebraic equations: Roots and implications. *Focus on Learning Problems in Mathematics, 5*(3, 4), 63-78.

Kieran, C. (1979). Children's operational thinking within the context of bracketing and order of operations. In D. Tall (Ed.), *Proceedings of the Third International Conference for the Psychology of Mathematics Education* (pp. 128-133). Coventry, England: University of Warwick, Mathematics Education Research Centre.

Kieran, C. (1981). Concepts associated with the equality symbol. *Educational Studies in Mathematics, 12,* 317-326.

Lapointe, A., Mead, N., & Phillips, G. (1989). *A world of differences: An international assessment of mathematics and science*. Princeton, NJ: Educational Testing Service.

Larkin, J. (1977). *Problem solving in physics*. Berkeley: Group in Science and Mathematics Education and Department of Physics, University of California at Berkeley.

Lindquist, M. (Ed.). (1989). *Results form the fourth mathematics assessment of the national assessment of educational progress*. Reston, VA: National Council of Teachers of Mathematics.

Lochhead, J., & Mestre, J. (1988). From words to algebra: Mending misconceptions. In A. Coxford (Ed.), *The ideas of algebra, K-12* (1988 Yearbook of the National Council of Teachers of Mathematics) (pp. 127-135). Reston, VA: National Council of Teachers of Mathematics.

Mathematical Sciences Education Board. (1990). *Reshaping school mathematics: A philosophy and framework for curriculum*. Washington, DC: National Academy Press.

McKnight, C., Crosswhite, F. J., Dossey, J., Kifer, E., Swafford, J., Travers, K., & Cooney, T. (1987). *The underachieving curriculum: Assessing U.S. school mathematics from an international perspective*. Champaign, IL: Stipes.

National Council of Teachers of Mathematics. (1989). *Curriculum and evaluation standards for school mathematics*. Reston, VA: National Council of Teachers of Mathematics.

Pimm, D. (1987). *Speaking mathematically: Communication in mathematics classrooms*. New York: Routledge.

Posner, G., Strike, K., Hewson, P., & Gertzog, W. (1982). Accommodation of a scientific conception: Toward a theory of conceptual change. *Science Education, 66,* 211-227.

Rector, J., & Ferrini-Mundy, J. (1986). Formal mathematics study and teacher beliefs and conceptions: Interactions and influences. *Proceedings of the Eighth Annual Meeting of the North American Chapter of the International Group for the Psychology of Mathematics Education* (pp. 256-261). East Lansing: Michigan State University.

Saxon, J. (1987). Why Saxon elementary books say no to calculators (Advertisement). *Arithmetic Teacher*, *34*(6), 20.
Shulman, L. (1986). Those who understand: Knowledge growth in teaching. *Educational Researcher*, *15*(2), 4-14.
Skemp, R. (1978). Relational understanding and instrumental understanding. *Arithmetic Teacher*, *26*(3), 9-15.
Sowder, L. (1986). Non-conservation of operation in American algebra students. In G. Lappan & R. Even (Eds.), *Proceedings of the Eighth Annual Meeting of the North American Chapter of the International Group for the Psychology of Mathematics Education* (pp. 90-93). East Lansing: Michigan State University.
Stevenson, H., Lee, S., & Stigler, J. (1986). Mathematics achievement of Chinese, Japanese, and American children. *Science*, *231*, 696-699.
Swan, M. (1983). *Teaching decimal place value: A comparative study of "conflict" and "positive only" approaches*. Nottingham, England: Shell Centre for Mathematical Education, University of Nottingham.
Tall, D. (1990). Inconsistencies in the learning of calculus and analysis. *Focus on Learning Problems in Mathematics*, *12*(3, 4), 49-63.
Tall, D., & Vinner, S. (1981). Concept image and concept definition in mathematics with particular reference to limits and continuity. *Educational Studies in Mathematics*, *12*, 151-169.
Tirosh, D., & Graeber, A. O. (1989). Preservice elementary teachers' explicit beliefs about multpilication and division. *Educational Studies in Mathematics*, *20*(1), 79-96.
Underhill, R. (1988). Focus on teacher education in diagnostic and prescriptive mathematics: Mathematics learners' beliefs—A review. *Focus on Learning Problems in Mathematics*, 10(1), 55-69.
Usiskin, Z. (1982). *van Hiele levels and achievement in secondary school geometry*. Chicago: University of Chicago, Department of Education. (ERIC Document Reproduction Service No. SE 038813)
Wagner, S. (1977). *Conservation of equation, conservation of function, and their relationship to formal operational thinking*. Doctoral dissertation, New York University.
Wearne, D., & Hiebert, J. (1988). A cognitive approach to meaningful mathematics instruction: Testing a local theory. *Journal for Research in Mathematics Education*, *19*, 371-389.
Wheatley, G. H. (1984). The importance of beliefs and expectations in the problem solving performance of sixth grade pupils. In J. Moser (Ed.), *Proceedings of the Sixth Annual Meeting of the North American Chapter of the International Group for the Psychgology of Mathematics Education* (pp. 141-146). Madison: University of Wisconsin.
Williams, S. (1990). The understanding of limit: Three perspectives. In G. Booker, P. Cobb, & T. Mendicuti (Eds.), *Proceedings of the Fourteenth Conference for the Psychology of Mathematics Education* (pp. 101-108). Mexico.
Willoughby, S., Bereiter, C., Hilton, P., & Rubenstein, J. (1981). *Real Math* (Teacher's Guide, Level 3). La Salle, IL: Open Court.
Wilson, P. (1990). Inconsistent ideas related to definitions and examples. *Focus on Learning Problems in Mathematics*, *12* (3, 4), 31-47.
Young, S. C. (1982). The mental representation of geometrical knowledge. *Journal of Mathematical Behavior*, *3*(2), 123-144.
Zykova, V. I. (1969). The psychology of sixth grade pupils' mastery of geometric concepts. In J. Kilpatrick & I. Wirzup (Eds.), *Soviet studies in the psychology of learning and teaching mathematics* (Vol. 1, pp. 149-188). Chicago: University of Chicago, School Mathematics Study Group. (Original work published in 1954.)

CHAPTER TWELVE

Assessing Mathematical Thinking and Learning Potential

HERBERT P. GINSBURG SUSAN F. JACOBS
LUZ S. LOPEZ

Abstract

In response to widely recognized social concerns and stimulated by new research in cognitive psychology, teachers and researchers are beginning to recognize both the need and the potential for significant change in the teaching and assessment of school mathematics. At present, however, assessment is still dominated by standard tests, which are inadequate to the task of revealing what the teacher most needs to understand in order to plan instruction: the student's thinking processes and strategies, and the student's learning potential. Fortunately, cognitive psychology has provided insights into the measurement of thinking and learning that point the way to new approaches to classroom assessment. They include a variety of interview methods, self-report procedures, and techniques for assessing learning potential. We describe thinking-oriented approaches to assessment and show how they can be adapted for classroom use, allowing the teacher to create a rich and practical theory of the individual student. Moreover, their use fosters a classroom climate hospitable to the implementation of a "thinking curriculum" in which both teaching and assessment contribute to intellectual development. This curriculum is based on the assumption that the core of education is the fostering of thinking. If the focus is on thinking, then the subject of the mathematics curriculum is not mathematics as it is sometimes conceived, namely a collection of meaningless procedures learned by rote. Instead, the curriculum is concerned with thinking about mathematics. Mathematics education should involve helping students to think mathematically and even, at a meta level, to think about their mathematical thinking. This thinking

> *curriculum requires a constant focus on (and assessment of) thinking, not on mathematics as a finished, static product. Thinking assessment requires the thinking curriculum and the thinking curriculum requires thinking assessment.*

In this chapter we make three arguments concerning assessment in mathematics education:

1. Teachers need to assess students' thinking and learning potential, not just the rote memorization of number facts or ability to use calculational procedures in a mechanical fashion.
2. Researchers have already developed assessment methods of this new type, and they can be adapted for classroom use.
3. The resulting assessment of students' thinking transforms classroom practice, leading to the "thinking curriculum."

The Need for New Approaches to Assessment

In response to widely recognized social concerns and stimulated by new research in cognitive psychology, teachers and researchers are beginning to recognize both the need and the potential for significant change in the teaching and assessment of school mathematics.

It is evident that too few American students—especially black and Hispanic students—reach high levels of achievement in mathematics (Cocking & Chipman, 1988). Most students—white and middle-class included—are inadequately prepared for participation in an increasingly technologically complex society (Johnston & Packer, 1987; National Research Council, 1989). If we are to broaden the social base of mathematically capable students beyond a small elite, we must reexamine the purposes and nature of mathematics teaching and assessment.

Teaching students to get the right answer in the shortest possible time with the least possible amount of thinking is no longer a useful goal in mathematics. Calculators and computers can execute tedious calculations far more efficiently than can humans. The 1989 *Standards* of the National Council of Teachers of Mathematics (NCTM) urge instead that we teach students to think creatively in order to cope with open-ended real-world problems. They urge also that the teaching of mathematics as a thinking activity involve consideration of students' beliefs and attitudes. Students must learn not only to think in flexible ways but also to appreciate that mathematics can be a creative and exciting activity.

Instruction and assessment are inseparable. If our teaching stresses rote learning, then our tests must attempt to measure it. And if we test for rote learning, using the results to evaluate students, teachers, schools, or nations in a competitive setting, then our teachers inevitably will prepare their students for the test. Teaching to the test is obviously most destructive when the test measures trivialities unrelated or antithetical to desired curriculum goals. But to the extent that assessment attempts to measure interesting aspects of thinking and learning, it can facilitate instruction (Frederickson & Collins, 1989).

At present, assessment is still dominated by standardized tests focusing on rote learning and mechanical use of procedures—rather uninteresting products of the student's learning. These tests are inadequate to the task of revealing what the teacher most needs to understand in order to plan instruction: students' thinking processes and strategies, and their learning potential. Fortunately, cognitive psychology has provided insights into the measurement of thinking and learning pointing the way to new approaches to classroom assessment.

Several alternatives to norm-referenced standardized testing are compatible with the goals of the 1989 NCTM *Standards*. They include a range of interview methods, self-report procedures, and techniques for assessing learning potential. These thinking-oriented approaches to assessment allow the teacher, at all grade levels, to create a rich and practical theory of the individual student. Moreover, their use fosters a classroom climate hospitable to the implementation of a "thinking curriculum" in which both teaching and assessment contribute to intellectual development.

What Do We Want to Assess?

One of Piaget's most important legacies is a focus on children's thinking. He taught us to examine the processes of children's mental activity, not its products. Piaget himself came to this insight as a young man when he was engaged in administering a standardized IQ test to young children. He soon became fascinated with the reasons for children's answers, right or wrong. He was less interested in the answers themselves than in the sometimes marvelous and apparently strange processes of thinking that produced them. Just as Freud, Piaget's early inspiration, attempted to dig below the surface of the manifest dream to uncover the cognitive "dream-work" beneath it, so Piaget attempted to discover the hidden thinking responsible for the IQ test answers (Ginsburg & Opper, 1988, p. 3). Piaget's investigations led to the conclusions that children's thinking is often different from the adult's and that even a "wrong" answer may result from interesting—and indeed "logical" and sensible—thought processes.

For our purposes, the details of Piaget's findings, which after all do not concern thinking about academic subject matter, are less important than his general goal of uncovering and describing basic thought processes.

Adopting this goal, we must attempt to gain insight into the thinking that produces the student's everyday mathematical behavior in the classroom. In doing this, we must move beyond the particular tasks Piaget devised; our interest is in determining how the student accomplishes column addition or the equivalence of fractions, not in investigating the cognitive processes responsible for performance on Piaget's classic conservation or classification tasks.

Here is an example: Sara (S), a 7-year-old second-grader, was asked, "How much is three plus four?" She used her fingers, trying to hide her actions from the interviewer, and obtained the answer "seven." Then the interviewer (I), Karen Schlichting, asked Sara to explain how she got the answer.

> S: What you do is this trick: You have three and you know that it's in your head and then on your fingers you count four, five, six, seven. You only have to use four fingers and not all of them, like three of them and then four. So you can see on your fingers that it is four.

From this explanation, the interviewer learned a great deal concerning Sara's understanding of addition. Clearly Sara did not produce the answer by rote memory; instead, she used the strategy of counting on from the first number. In this strategy, the first number can be represented mentally, the second physically, by the fingers, and the sum by the counting words. She knew that number combination questions can be solved by counting, but she did not yet realize that it would be easier to count onward starting from the larger number. Sara's answer reveals a good deal of complexity behind her response to one of the simplest aspects of arithmetic, a small number combination problem.

Later, Sara was given a simple verbal subtraction problem, "How much is two take away one?" She answered, "Two take away one equals one."

> *I*: What does "equals" mean?

> *S*: Equals means that you are giving your problem an answer. If you have a math problem, say three take away two, and that's very easy because the answer is one, but if you just put one there, the teacher would think your answer is 21. [She wrote: 3 − 2 1.] So you need to put some sign to say that these numbers equals the answer.

> *I*: Is there another word you could put there?

> *S*: No, equals is a sign in math, and you can't put a word in there or the teacher would think you didn't understand that you are doing math.

This exchange reveals a good deal about Sara's understanding of the = sign. First, she does not seem to think of = as indicating equivalence;

instead, she has an "action interpretation" in which = indicates that the student is supposed to get an answer to the stated problem. Moreover, an important function of the = sign is to separate the written numerals so that one can distinguish between the second addend and the sum. Also, the = sign is part of a distinctive language, the language of mathematics, which must be spoken in class and which is not readily translatable into English. This, then, is what we mean by children's thinking: not just the right or wrong answers to the various problems, but reasoning, beliefs, and theories of the sort displayed by Sara. The goal of assessment should be to discover and understand thinking in order to foster students' learning and to help them overcome their difficulties.

How Can We Assess Mental Processes?

How can we gain insight into student's mental processes? It is unlikely that we can learn about these interesting features of understanding simply by recording and evaluating the correctness of a student's answers to standard test problems like 3 − 2. Consider other types of assessment that may be useful: the clinical interview method and its derivatives, self-report methods, and techniques for assessing learning potential.

The Clinical Interview and Its Derivatives

The essence of standardized testing is the constraint that the interviewer administer the test to all children in as constant and unvarying a manner as possible. This means asking all children the same questions, using the same words. Under conditions of standard administration, the tester is not allowed to rephrase the question if the child does not seem to understand it, and the tester is certainly not allowed to pose different questions to different children. After all, it would be unfair to treat some children differently from others.

In the course of his work with IQ tests, Piaget concluded that standardized testing—despite its laudable goal of treating all children fairly—often provides little insight into children's thinking. The main weakness of standardized testing, according to Piaget, is its very essence (and in a sense its strength)—its requirement that all children be treated in the same manner. If this is done, then the examiner cannot effectively deal with misunderstanding of questions and cannot probe for methods of solution. For example, if a child speaking a dialect fails to understand a key word, the examiner cannot reword the question, in dialect, more effectively to tap the child's understanding. Or if the hypothesis is that the child's response is produced by a particular strategy, as in the case of Sara's counting on method using her fingers, then the examiner is constrained from devising new questions to identify that strategy more definitively.

Given these weaknesses inherent in standard testing, Piaget (1929) developed the "clinical interview" method,[1] which we term "flexible interviewing" and which we believe is the most informative—and difficult—assessment method available for the purpose of assessing mathematical thinking.

Flexible Interviewing

The essence of this kind of interview is its flexible, responsive, and open-ended nature. Although at the outset the interviewer has available several tasks likely to be appropriate for the topic at hand, initial questions are intentionally quite general, allowing the student to determine the direction and content of the interview. Specific tasks and questions presented to the student as the interview progresses are determined by the student's responses to the initial questions and tasks.

The flexible interview is a dialogue in which the interviewer asks the student to reflect on and articulate thinking processes. A flexible interview is time consuming and demanding, requiring twenty minutes to an hour of concentrated effort. A good interviewer will have command of relevant mathematical content and must be familiar with typical mathematical thinking at the student's level.

The interviewer is responsible for establishing rapport, preparing a series of appropriate tasks, and listening to and observing the student's responses. The interviewer's goal is open-ended. Topics of inquiry and issues under investigation may not have been planned or even anticipated prior to the interview. Because they have been suggested by the student's responses to initial questions, the issues investigated are likely to be of immediate interest or concern to the student. The observant and thoughtful interviewer constantly adapts interviewing strategy to the direction suggested by the student.

Nonspecific questions ("How did you do it?" "What did you say to yourself?" "How would you explain it to a friend?") are especially useful because they encourage rich verbalization but give no suggestion of how the student should respond. Students should be encouraged to work with chips or tiles, pencils or marker pens and paper, and small toys. These manipulative objects motivate the student and, because they tend to externalize thinking, allow the interviewer-observer additional insight into thinking processes.

Responding to the student's responses and behavior, the interviewer makes and tests hypotheses about the student's thinking. Tasks are varied and modified, becoming more specific in order to focus on particular aspects of thinking, and more difficult in order to test the limits of the student's understanding.

Flexible interviewing is nonstandardized in many ways. Establishing rapport requires warm encouragement and calming of apprehension for some students. For others, a tone of high expectation produces better results. Whereas one student can be allowed to follow a train of thought at length, another student will need closer control, with more frequent questions. Questions must be phrased so that the student understands them. A skilled interviewer is aware of the student's personal vocabulary and can rephrase questions in the student's words.

The interviewer is not satisfied with a "correct" answer but tests the strength and consistency of the student's beliefs by using repetition and countersuggestion. An "incorrect" answer may reveal valuable information about thinking processes. Whether "correct" or "incorrect," the most desirable answer is one that provides the most information about the student's thinking.

Obviously, maintaining the student's interest is necessary if the student's best, rather than ordinary, performance is to be measured. The bored, distracted, tired, or uncomfortable student will not reveal much useful information. When the student is interested in the issues, enjoying the attention of the interviewer, and when the level of difficulty of the task is challenging but not overwhelming, a rewarding dialogue results. The interviewer may learn not only about the student's thinking, but also about important attitudes and beliefs concerning the student's own ability and concerning the goals, methods, and nature of mathematics.

Consider two examples. The first shows that a correct response may lie on a foundation of unsound understanding. The interviewer observed that Chris (C) wrote the correct answer, 14, when given 9 + 5.

I: Now I'm going to ask you something about 14. How come you wrote 14 with a 1 and then a 4?

C: 'Cause that's how I write 14.

I: I notice that when you write 14 you have a 1, and on the right of that is a 4. What does that 4 stand for?

C: 'Cause it's 14.

I: All right. What does the 1 stand for?

C: That's how you write 14.

I: Why don't you write it like this [41]?

C: That's 41.

I: All right. Why do you write 41 like that?

C: Because there's a 4 there and a 1 there.

I: Why did they invent that way of doing it? What could it possibly mean? What does that 1 stand for?

C: 1.

I: What does that 4 stand for?

C: 4.

I: Can you write the number 123? That's right. That's how people write 123. What does that 1 mean?

C: 1.

I: Just 1. And what does that 2 mean? What does it stand for? What is it telling us?

C: 2.

I: Just 2. And the 3? What does that tell us?

C: Just 3.

Clearly, Chris did not understand the place value meaning of the symbols he wrote accurately.

The second example shows that the child's wrong answer may be based on sound understanding.

I: Rabbit has seven carrots and he gives two of them to squirrel. How many carrots does rabbit have left?
Becky (B): Four.

I: How did you figure that out?

B: Well, I knew that four and two is seven, and if rabbit has seven carrots and he wants to share them with squirrel so he gives squirrel two, he must have four left.

Becky's incorrect response was based on a rather subtle pattern of reasoning involving a syllogism. Her logic was that if it is true that four and two is seven, then it must be true that seven take away two is four. Her reasoning is flawless; her premise is wrong. For the teacher, knowing about her interesting pattern of thought is more valuable than merely noting the fact that her answer is wrong.

The Structured Interview

Though powerful, the flexible interview is extremely difficult to administer. It requires impromptu theorizing and creation of questions designed to provide critical tests of hypotheses, all of which is to be done while the examiner is attempting at the same time to be sensitive to the child's emotional needs. To reduce some of these pressures, researchers have sometimes found it useful to employ a less flexible approach, the "structured interview."

The salient features of this method are its semistandardization and the fact that the questions are not entirely contingent on the student's responses. The same problems and the same questions are posed to all students. Students' answers or explanations are recorded as such, without channeling the student's activity toward a particular area. The interviewer does not test new hypotheses as they arise. Thus, the structured method is basically a standardized interview in which some of the questions may be open-ended and some multiple-choice. The method requests reflection, asking students to tell everything they were thinking while implementing a given process.

The method has proved useful when employed in an area of mathematical thinking that has been extensively explored, so that students' processes and strategies are familiar and easily described. The major limitation of the method is that the examiner's flexibility of questioning is limited. Because the examiner cannot ask questions contingent on the student's answer, the amount of information obtained may be limited. The examiner may be prevented, for example, from following up on a student's apparent misunderstanding of a key question so as to uncover a level of understanding that was not immediately obvious.

In the structured interview, the examiner usually uses a few simple and noncontingent questions, like "How did you solve the problem?" and "Tell me everything you were thinking." Often these yield considerable insight into children's methods of solution.

For example, Laura was given this word problem:

Sue has 5 crayons. She has 3 fewer crayons than Willy. How many crayons do they have altogether?"

Once Laura had solved the problem, the following interview was conducted:

I: Tell me everything you were thinking.

Laura (L): Once I added the three and the five, I stopped, thought about the question they asked me in the problem, and decided it was correct.

I: Tell me more.

L: I thought I did have to add those two numbers, since Sue had three fewer than Willy. Fewer meant that he had more and when we add we get more, so I knew it was correct.

I: Tell me everything you were thinking.

L: When I read the problem the first time I wasn't sure if Willy had 5 or if Sue had 5.

I: Tell me more.

L: I reread the problem. Once I reread, I knew Sue had 5 because the problem said it.

The interview shows that Laura monitored her problem-solving performance when she checked to see if the operation she had selected was the appropriate one. We also see that she was able to evaluate her understanding of the problem and used rereading as the strategy to reduce her confusion.

Organized Probes

Another, even more structured variation on clinical interviewing is the "organized probe." Ginsburg (1990) has developed a comprehensive series of organized probes in connection with the Test of Early Mathematics Ability or TEMA (Ginsburg & Baroody, 1990), which focuses on various aspects of informal and formal mathematical knowledge in children from preschool through approximately grade 3. The test items were selected from data-gathering procedures used in cognitive developmental research on mathematical thinking over the past twenty or thirty years. The test items deal with such matters as informal addition, the mental number line, the concept of cardinality, simple number combinations, alignment procedures for addition and subtraction, and base ten concepts.

Although the TEMA covers interesting areas of mathematics performance, it does not shed a great deal of light on thinking processes generally or on understanding in particular. Consequently, Ginsburg (1990) developed an organized system of probes to be used in parallel with the TEMA. The general idea was that after the TEMA had been given in the standard fashion, many examiners would find it useful to probe further into the thought processes that produced the observed performance, particularly in the case of errors. Most evaluators, however, have not had training or experience in assessing students' thinking. Consequently, Ginsburg attempted to provide examiners with a structured and comfortable procedure for probing the strategies and concepts underlying students' responses to the TEMA. Of course, clinical interviewing is a more effective and difficult means of achieving the same end; yet because most evaluators are not prepared to engage in extensive clinical interviewing, organized probes are a useful first step.

The probes for each of the 65 items of the TEMA first attempt to establish whether the student has understood the basic question. Often students produce an incorrect response because they have misinterpreted a minor feature of the question asked. The probes attempt to distinguish this situation from that in which students do not understand the question because they fail to comprehend the relevant concept.

For example, one of the TEMA items attempts to deal with the addition of multiples of ten by asking, "Here are some questions about adding money. We'll pretend that you have some money and I give you some more. If you start with nine dollars and I give you one ten dollar bill, what do you end up with?" The wording was, of course, intended to be clear in the first place, but some children misunderstand the question, perhaps through from confusion about the phrase, "What do you end up with?" In a true clinical interview, the examiner would attempt to determine what each child failed to understand and then reword the question accordingly. In the system of probes, on the other hand, the examiner has available two alternative questions. One attempts to deal with the "What do you end up with?" problem by substituting the following: "If you have nine dollars and I give you one ten dollar bill, how much do you have altogether?" A second question is intended for children who may be confused about money: "If you start with a nine, and then you add on a ten, how much do you end up with?" Given such alternative questions, some children suddenly see the light and exclaim, "Oh, is that what you meant!"[2]

Next, the probes attempt to determine the strategies and processes used by the student to solve the problem. For example, in the case of concrete addition, the basic TEMA question is: "Joey has two pennies. He gets one more penny. How many does he have altogether? If you want, you can use your fingers to help you find the answer." As the questions are asked, the examiner shows the child the numbers of pennies involved. The probes then attempt to determine whether the student used such procedures as counting on the fingers, mental counting on, or memorized number facts. Three types of questions are suggested. The first is "It's O.K. to use your fingers. Put your hands on the table and show me how you do this. Tell me out loud what you are doing." In this case, some children who had previously thought that it is improper or even cheating to use their fingers breathe a sigh of relief and show the examiner their finger-counting procedure. A second question is: "How did you know the answer was _____?" In this case, some children say, "I know that three and two is five because I just knew the answer. I learned it." If the answer was given quickly, this may be a good indication of the use of memorized facts. A third question is: "Why does six pennies and two make _____ altogether?" In response to this question, some children say, "It had to be eight because I know that six and one is seven and then one more must be eight." Here the use of a reasoning strategy is evident.

Summary

The flexible interview is deliberately nonstandardized: The examiner poses questions sensitive to the individual's sometimes unique approach and, on the spot, constantly tests hypotheses concerning thought processes that might account for both correct and incorrect answers. Though powerful, the flexible interview is hard to conduct; indeed, Piaget claimed that it required a year's daily practice to acquire skill in clinical interviewing. Derivatives of the flexible interview have been developed to make it more accessible to researchers and diagnosticians. The structured interview is relatively standardized, using key questions without follow-up. The system of probes involves a series of specific questions designed to clarify individual standard test items and uncover the thought processes used in their solution.

Self-Reports

The use of self-report procedures has proliferated in the past few years. The most commonly used self-report procedures are thinking aloud and retrospective reports (Ericsson & Simon, 1980).

Thinking Aloud

In the thinking-aloud procedure, the student is asked to "talk out loud" or to say everything he or she is thinking while solving the problem. It is essentially an "on-line" or concurrent verbalization. The examiner does not call for explanations, elaborations, or retrospections of what the student is doing.

This method has proved to be an effective way to assess the cognitive and metacognitive processes used during problem solving. As a student verbalizes his or her thinking, the examiner can identify the strategies and processes the student is using to carry out the task. Sophisticated methods to code self-report data in the area of problem solving have been derived from Polya's work (1945). Exemplars of these coding systems can be found in Schoenfeld (1983, 1985) and in Artzt and Armour-Thomas (1990).

Thinking-aloud has been used widely, particularly to supplement standardized tests. For example, the Diagnostic Test of Arithmetic Strategies (Ginsburg & Mathews, 1984) uses this procedure to investigate students' invented strategies. This technique has also been used with considerable success in the field of problem solving (e.g., Artzt & Armour-Thomas, 1990; Newell and Simon, 1972; Schoenfeld, 1983; Swanson, 1991).

To illustrate the use of this procedure, we present Peggy's (P) thinking-aloud protocol as she solved an arithmetic word problem:

> *I*: I am interested in how children think. I am going to ask you to solve a problem and to think out loud while you are solving it. Tell me everything you are thinking as you are thinking about it. Here is the problem: Dave bought a bag of 12 pieces of animal food. His father

gave him some extra pieces. Now he has 16 pieces of animal food. He dropped two of the pieces of animal food his father gave him. How many of the pieces of animal food his father gave him did he have left?

P: I am thinking that he gave, he bought a bag of 12 pieces and his father gave him some more pieces. It does not say how many more. But it says here [rereads], 16 pieces. So, I am thinking that if it was 12 first and then 16, you have to add it because 12 is a lower number than 16. So, 12 [writes 16, silence], 28. [rereads] So, now he has 16. But, I am thinking that if he dropped two and he had 16 pieces, you have to take away 16 from two, because he dropped two. 16 take away two is four [writes]. And, I am thinking that now he would have again 12 pieces [rereads]. Oh, no . . . I am thinking that he has 14 pieces of animal food because he dropped two. [rereads] I am thinking I have to find out how much animal food his father gave him. 16 take away 12 is 4. His father gave him four pieces and then, he dropped two. So now I take away two. So, now he has two pieces left.

An analysis of Peggy's protocol makes apparent the use of the following processes. After reading the problem, Peggy stated the data and the unknown. As she reread, she stated the data again, made a plan to add, and implemented her plan. Peggy reread again, restated the data, made another plan, this time to subtract, and implemented it. As she reread once more, she monitored her performance and subtracted again, this time correctly. The last time she reread, she stated the goal (to find out how many pieces of animal food Dave's father gave him), implemented her plan, and carried out the calculation to solve the first step of the problem. Next, she stated the data, verbalized a plan, and implemented it.

Retrospective Reports

In the retrospective report procedure, the student is asked to verbalize his or her thinking immediately after the completion of the task rather than during its execution. From this information, the examiner can identify and code the strategies and processes the student used to solve the problem. Retrospective reports have been used often in the field of problem solving (e.g., Schoenfeld, 1983, 1985). Our experience with retrospective reports has shown that they foster students' reflection. Often, while giving a retrospective report, students realize their mistakes and correct their answers.

To illustrate the use of this procedure, we present Peggy's retrospective report on the same problem concerning which she had done the thinking aloud described above.

I: Tell me everything you were thinking while you were solving the problem. Make believe I did not hear you solving the problem and you are now telling me everything you were thinking.

P: I was confused so I reread. I was thinking that he had 12 pieces of food. Then I read some more and it said his father gave him some more pieces of animal food. And then I read more and it said that now he has 16. So I thought if 12 [silence] is smaller I'll have to add some more because I can't take away 16 from 12. I added and I thought it was wrong. So I read again to check. Then I subtracted it and found out that his father gave him four pieces. And then I read he dropped two of them. I thought I had to subtract because he dropped them, he didn't have them. So then I subtracted four from two and he had to have two left.

From Peggy's retrospective report we learn about some of the processes she used. It is apparent that she monitored her understanding and used rereading as the strategy to help solve her confusion. She restated the data, expressed a plan, implemented it, and monitored her problem-solving process. As she monitored, she again used rereading as the strategy to help her check the solution process. Next, she implemented a new plan. Finally, she stated the data, planned again, and implemented the second step to solve the problem.[3] Contrasting Peggy's retrospective report and her thinking-aloud protocol shows that both procedures provide valuable information regarding the processes Peggy used; from neither procedure alone can we obtain a full picture. The thinking-aloud protocol illustrates the diverse processes she used while problem solving. The retrospective report makes clear the metacognitive skills she used: She monitored her performance and used rereading as the strategy to help her reduce confusion. This information was not evident from the thinking-aloud protocol. Thus, the two procedures may complement each other.

Assessment of Learning Potential

Several approaches to assessment have in common the goal of assessing a student's potential for learning. Research in this area evaluates learning potential on the basis of the amount and the kind of intervention required for success. Intervention may include graded hints and prompts, dialogue, guided group discussion, and teacher modeling of processes and strategies. Presumably, the less direct intervention required, the more potential for independent learning the child possesses. This approach can also suggest how much and what kind of help is best suited to the individual student, and in this sense suggests teaching methods.

Vygotsky (1978) and Feuerstein (1979), stressing the social aspects of learning and cognitive development, have exercised major influence on the assessment of learning potential or, in Feuerstein's terms, "dynamic assessment." At the core of Vygotsky's theory is the concept of the "zone of proximal development," defined as "the distance between the actual developmental level, as determined by independent problem-solving, and the level of potential development, as determined through problem-solving under adult guidance, or in collaboration with more capable peers" (Vygotsky, 1978, p. 86). In the United States, several approaches have developed from Vygotsky's theory of social learning, including the work of Budoff (1987), Carlson and Widaman (1986), and Ann Brown and her colleagues (e.g., Campione, Brown, Ferrara, & Bryant, 1984).

The agent of cognitive growth to which Feuerstein attributes the most importance is "mediated learning," the training given by a teacher, peer, parent, or other social agent, who frames, selects, focuses, and feeds back an environmental experience in such a way as to create the opportunity for appropriate learning.

We have begun some work using the approach of Brown and her colleagues to assess learning potential. First, a pretest is given to the student. Next, an analysis of the student's mistakes suggests to the examiner the areas in which learning potential can be assessed. Following this, the examiner presents the student with a problem in an area in which he or she has shown difficulty and provides hints to assist in the problem-solving process. These range from general metacognitive hints to those specific to the demands of the task. The amount of help the student needs is an estimate of learning efficiency within that domain. The examiner continues presenting the student with problems similar in nature to the initial one, providing as much help as needed, until the student is able to solve problems independently. After helping the student to achieve independent learning, the examiner presents near, far, and very far transfer problems and students are given assistance as needed to solve them. In some cases the students are given a posttest to assess the gains following the assessment.

We present an example involving Jamel, a fourth-grade student who was having difficulties in mathematics. First, an assessment of Jamel's arithmetic skills indicated that Jamel had solid informal mathematical skills but that lack of number fact knowledge was hindering his performance. Therefore, we conducted an assessment of his learning potential for number facts. The following dialogue took place between Jamel and the interviewer after several months of work together. At the time of the dialogue, Jamel had already learned to interpret multiplication as repeated addition.

I: What is 9×3?

J: [No answer]

I: Can you think of a way to figure out what 9 × 3 is?

J: [No answer]

I: 9 × 3 is the same as adding a number several times.

J: 9 + 9 = 18 [uses fingers to count].

I: So . . . 18 is adding how many 9's?

J: Two.

I: So . . . 9 + 9 would be the same as saying 9 × 2. Does this give you any ideas about how to continue?

J: I don't know.

I: How many nines are we interested in finding the sum of?

J: Three.

I: Does this give you any ideas about how to continue?

J: I have to add one more nine [uses fingers and adds 18 + 9 = 27 by counting on from 18].

I: What does that 27 mean? Where does it come from?

J: Nine times three is 27.

I: How would you know that 27 is the correct answer?

J: [No answer]

I: Can you prove in another way that 9 × 3 is 27?

J: [No answer]

I: How about doing it with tallies?

J: [Right away makes 3 sets of 9 tallies each and then counts them out loud to equal 27.]

I: Is there any other way to check it?

J: [Writes: 9 + 9 = 18; 18 + 9 = 27.]

I: Great job!

J: [Smiles]

The assessment showed that Jamel needed several hints and prompts to access a strategy that could help him find the answer to 9 × 3. However, he did profit from the prompts. Accessing strategies to check his work was also difficult for Jamel. With assistance, he accessed even another checking strategy on his own. Further, Jamel showed the ability to connect his informal math knowledge to his formal math knowledge quite readily, as he was able spontaneously to represent the numbers with tallies.

We see then that although Jamel had been unable to learn these facts in any one of the ways he had tried previously (including drill), several prompts helped him to acquire useful thinking strategies, to develop understanding of the facts and the ability to transfer his knowledge. Note that the prompts are based on the assumption that internalizing basic number combinations can be facilitated by recognizing and exploiting relationships. With many children, even students with mathematical learning difficulties, this meaningful approach is more likely than drill to promote number fact mastery and to foster analytic thinking and problem-solving skill.

As in the case of flexible interviewing, the assessment of learning potential can be enormously informative but very difficult to conduct. To reduce the difficulty, especially for novice assessors, Ginsburg (1990) has developed a series of organized probes to assess learning potential. The issue is whether the student can learn the type of material covered in the TEMA (Ginsburg & Baroody, 1990) with a minimum of hints or whether more substantial teaching is required. In the first case, it is clear that the student is close to "understanding;" in the second case, the student is not.

For example, one TEMA item assesses the child's understanding of the "mental number line" by posing such questions as, "Which is just a little different from seven, one or nine?" The learning potential probes are of two types. First the examiner provides a hint: "Try to do it by counting. See if that will help." If this minimal amount of help is successful, clearly the child's potential for learning about this concept is substantial; indeed, the concept may be considered to have been on the tip of the "mental tongue." If the hint is not successful, the examiner goes on to offer "teaching," a much more directive approach: "You can tell how close they are by counting. First, you count from seven to nine, like this, 'eight, nine.' That's two counts. Then you count from one to seven. 'Two, three, four, five, six, seven.' That's six counts, so one is far away from seven and nine is close."[4] If the child then succeeds with this kind of help, there is again some potential for learning. If the child does not, learning potential—for this task, at this time—may be severely limited.

Assessment in the Classroom

Many would agree that the assessment methods described here are powerful. But can they be used effectively by ordinary teachers? Or, more precisely, can teachers who put the effort into learning the methods make practical use of them in the hurly burly of the everyday classroom?

The assessment methods described above are most effective when student and teacher work together in a one-to-one relationship. Although time constraints prevent many such occasions in the classroom setting, the teacher can use variations of the assessment techniques to focus attention on the student as individual, even within the context of a large group. But the assessment of the individual's thinking can occur only in classrooms in which thinking is encouraged. The traditional teacher who instructs students in standard algorithms and attends mainly to their correct and incorrect answers is not in a position to learn a great deal about understanding. By contrast, teachers who encourage students to engage in mathematical activities, to develop their own methods of solution, to discuss mathematical ideas and procedures, and to believe more generally that their own approaches to learning are valued can relatively easily learn a great deal about students' understanding. Putting it more bluntly, the teacher who does not encourage understanding cannot measure it in the classroom; the teacher who encourages it can learn a great deal about it. The teacher who always talks *at* students will learn little about their thinking; the teacher who talks *with* students can learn a good deal about how they solve problems.

The main differences between the "pure" assessment methods and variations suitable for classroom use are considerations of scale and of privacy. With regard to scale, time limitations prevent classroom interviews from being as extensive as research interviews; nothing can be done about this. With regard to privacy, the teacher's responsibility to deal with the class as a whole sometimes requires that the examination of the individual student's thinking be conducted in public. The loss of privacy need not be intrusive or upsetting if students' responses are accepted by the teacher and other students in a nonjudgmental way. Other students can even learn indirectly from a thinking assessment. Indeed, Stigler (1988) has shown that public examination of thinking can be conducted to everyone's benefit: In Japanese classrooms, children experiencing difficulty are often asked to present their methods of solution before the entire class, who then take a constructive role in discussion and analysis. If the exercise is conducted in a thoughtful and caring fashion, should our children be more concerned than the Japanese children with "losing face"? Is privacy more important than the ethic of shared learning?

Thinking assessments can take several forms in the classroom. Although most applications have been targeted to the elementary level, with suitable modifications they can be used at older age levels as well.

Interview Methods

Several researchers have advocated the use of mini-interviews in the classroom. These are usually, but need not be, structured interviews. Peck,

Jencks and Connell (1989) reported that they were able to interview 32 students in one hour every five to six weeks at the beginning of a new topic of study. Dionne and Fitzback-Labrecque (1989), in research involving administration of 10- to 15-minute interviews by a mathematics specialist, uncovered difficulties hidden by correct responses on tests, and provided diagnostic information helpful to the teacher and the students' parents. Recently, a major textbook publisher has made available a series of flexible interview activities designed to be used in connection with its texts (Silver Burdett & Ginn, 1991).

We speculate that students themselves may learn to act as structured and even flexible interviewers. One student may interview the teacher, or another student. In the style of a journalistic interview, the student's goal may be to write a protocol of the other person's thinking process while doing an arithmetic problem. That person will be able to judge how adequately the strategy was represented. Or one student may pretend to be a famous mathematician who has just discovered, for example, a method for doing two-digit multiplication. Another student may record an interview with this celebrity. Or students in a group might compare different methods of solution they have discovered through interviews: Which is the "best" or the "most interesting" or the "neatest"? Students may use interviews to obtain descriptions of mathematics-related beliefs and attitudes. Does mathematics always involve getting a correct answer? Is there always a correct answer? The results of these polls may be written up in a class mathematics journal.

Students in the middle grades very much enjoy making flow diagrams of ordinary procedures, such as washing their hands or making a purchase. After such experiences, they might attempt a flow diagram of a simple arithmetic procedure as revealed by a flexible or structured interview. One student may record steps in a procedure as it is performed by another student. Even the simplest procedures can be complicated, so it is important to start with a well-understood procedure at first. After the flow diagram is made, the steps it describes should be performed in an ongoing attempt to correct and adjust inaccuracies.

Following the example of the TEMA probes, teachers can devise follow-up questions relating to a quiz or to written work done in school or at home. For example, after selected homework problems, teachers may require students to respond to questions like "How did you get the answer?" and "How can you make sure that $2 \times 6 = 12$?" A few good questions of this type may yield answers suggesting subsequent teaching or reteaching of content that was poorly understood.

Self-Report

Thinking aloud and retrospective methods suggest a sharing of strategies for working with an arithmetic problem. The teacher emphasizes that there may be many successful strategies and elicits as many as possible as the students are solving various problems, sometimes in front of the group as a whole. Students can then discuss the strategies thus revealed, attempting to predict which will be successful and how others need to be modified.

Here is an example concerning one second-grade teacher—Joan Gansfuss—who encourages analysis of thinking in her classroom. The teacher wrote a simple computational problem, $9 + 7 = $ _____, on a large piece of paper in front of the class. She asked the students to solve the problem in their own way and to write down the answer. After this had been done, the teacher spent a good part of the "math lesson" exploring the students' strategies, like:

> I took 2 away from the 9 and that was 7. $7 + 7 = 14$. I add 2 more and I got 16.
>
> First I took the 7 and then I put up 9 fingers and I counted up 7 8 9 10 11 12 13 14 15 16.
>
> I knew $10 + 7 = 17$, but 9 is 1 less than 10, so 1 less than that $= 16$.

She asked all of the students in the room to explain their method of solution. Sometimes she encouraged them to describe it in writing. In the course of an exercise like this, students can provide very clear information concerning different strategies employed. With our assistance, the teacher had developed a simple scheme for coding the observed strategies. This scheme includes simple descriptions of procedures commonly observed in the research literature; recall of number facts; concrete counting involving fingers or other easily available objects; mental counting procedures like counting on; and various regrouping strategies, like "6 and 4 is 10 because I know that 4 and 4 is 8 and the answer is only 2 more than that." It is relatively easy for the teacher to record students' use of these strategies on a simple checklist, which provides a convenient record of the strategies the students were using. This is a clear improvement over the usual procedure in which the teacher simply attempts to remember the methods used by various students.

Learning Potential

As far as we know, little has been done to modify dynamic assessment techniques for use in the classroom. Here the focus should be not so much on discovering what the student has learned, but on the ability to learn new kinds of material and on the method of learning it. Of course, many good teachers use informal dynamic assessment methods to help students move along a continuum from well-known to new but closely related material. Their efforts may serve as models for developing methods of dynamic assessment suitable for classroom use.

Toward the Thinking Curriculum

Clearly, the various assessment methods can be used in classrooms to obtain important information concerning students' thinking. But using them also affects students in important ways and transforms the way in which teaching is conducted.

How Thinking Assessment Affects Students

The focus on thinking can provide students with several valuable lessons. One defines the "curriculum."

Thinking Is Taught Here
The teacher's focus on the student's thinking says, in effect: "I value your thinking, your way of dealing with problems. I am interested in your thinking; it is the center of my attention. Your thinking is so important that I must understand it. Your thinking is so powerful that it determines how I teach; it influences what happens in the classroom. In this class, we do not just memorize material; we solve problems by thinking. We always use our minds and we also think about using our minds. Thinking is taught here."

Metacognition
Another lesson involves training in meta-cognition. As an example, consider the retrospective procedure described earlier. Recall that in a second-grade class, discussion centered on students' descriptions of their methods for solving simple computational problems. Throughout this process the teacher established an atmosphere in which students were encouraged to share their strategies, to value the range of strategies arising in the classroom, and to verbalize strategies as explicitly as possible. The teacher showed the class that there can be a variety of ways to obtain the correct answer. After exposing the students to a number of strategies, she asked them if they would like to choose a different strategy and, if so, why they would choose it. Some students preferred to continue in the strategy already chosen, whereas other students chose a new approach, often "because it is faster." In any event, although the assessment procedure provided the teacher with valuable information concerning students' thinking, it also provided the students with a lesson in what thinking is all about. In effect, the lesson involved the sharing, valuing, and teaching of thinking about thinking.

The assessment methods are interactive, requiring verbal communication. The thinking curriculum helps the students to express their thought, to put it in words. Expressing thought requires students to focus attention and reflect on their own problem-solving processes. The assessment methods help students to develop metacognitive skills used to monitor, check, and evaluate their performance. The general lesson is: "Metacognition is taught here."

Some Effects on Teachers

Recently one of the authors had the opportunity to train a group of teachers in the country of Colombia on how to use some of the thinking assessment techniques. Teachers were trained to use thinking-aloud procedures, retrospective reports, and structured clinical interviewing techniques to assess students' thinking. These techniques gave a new dimension to the teachers' work. The teachers used these assessment techniques in the classroom not only to investigate thinking, but also as tools to foster it.

For example, in a fourth-grade classroom, when a student was having difficulty solving a word problem, the teacher abandoned her usual strategy of solving the problem for the student. Instead, the teacher asked the student to think aloud. Thinking out loud fostered this student's reflection on his problem-solving process. He was able to monitor his performance and correct his misunderstanding.

Another teacher proceeded as follows with a student who was having difficulty solving another word problem. First, she asked the student to show her his work and noted that the student had chosen addition as the operation necessary to solve the problem. At this point, the teacher changed her approach. Instead of telling the student that addition was not the correct operation to use, she asked the student to tell her how he had figured out that he had to add. The teacher became aware that the student was being guided by the key word "altogether." It happened that "altogether" in this case was a misleading word, not signifying addition, as it often does. Next, the teacher helped the student reason about the problem, making more use of interviewing techniques.

Yet another teacher began to ask her students to write out how they had solved the problem immediately after they had solved it. This teacher was impressed with what she was learning from her students. Not only was she aware of their use of sophisticated strategies such as analysis and exploration, but she was able to see the sources of their misunderstandings.

The Teacher as Thinker

The thinking curriculum requires the teacher to think, too. To teach thinking, the teacher has to model thinking: introspection, self-monitoring and checking, and weighing of alternative strategies. At the elementary school level, the experienced mathematics teacher probably knows the material so well that thinking aloud, verbalizing weighing of strategies, and introspection are in a sense staged for the students in order to help them learn these metacognitive skills for themselves. When there is a real mathematical question in the teacher's mind, the process acquires greater urgency and greater potential for learning. Challenging questions may inspire curiosity and interest in follow-up work. Sometimes a student will offer a step toward solution of the problem. If the teacher respects this offer, even if it is imperfect or unclear, the students can learn to see the value of exploratory thought. As a thinking model, the teacher may admit uncertainty and ask the student to rephrase the question or in other ways try to elicit clarification. In this small but real-world situation, the students observe the methods their teacher uses to grapple with a problem.

It is important for students to know that adults do not have ready-made answers on all occasions. Many times during the day teachers have to make decisions and plans that affect their students. The teacher's weighing alternatives and thinking aloud gives students indirect experience in the decision-making or planning process. As students are encouraged to enter into this process with their own comments and suggestions, the learning experience becomes more direct. When students actually provide helpful, imaginative, and constructive ideas, the teacher and students benefit together.

Modeling is one of the first steps involved in good teaching. In Colombia, teachers modeled thinking aloud and giving retrospective reports before asking students to implement these methods in the classroom. Teachers reported that this experience made them aware of the amount of effort involved in the activities in which they were going to ask their students to engage.

The Teacher's Sense of Self as a Teacher

In Colombia, the assessment techniques fostered reflection in the teachers. Prior to our joint work, they viewed themselves as authority figures. Along with this view was the belief that they should present themselves to students as people who knew everything and did not make mistakes. As our work developed, teachers began to see the importance of presenting themselves to their students as human beings. Teachers even began to model confusion and making mistakes in the classroom.

This implied that the teachers' sense of the nature of mathematics education had changed. They downplayed their emphasis on right or wrong answers. Instead, they asked students to spell out the processes used to solve problems, and they gave more credit for this aspect of problem-solving than to the answer itself. The curriculum had shifted to mathematical thinking.

A Conclusion of Reciprocal Implication

Classroom assessment should focus on mathematical thinking. It is now possible to do this with some success, and the assessment of thinking leads to a distinctive approach to mathematics education (and education in general). This type of assessment is based on the assumption that the core of education is the fostering of thinking. If the focus is on thinking, then the subject of the mathematics curriculum is not mathematics as it is sometimes conceived—namely, a collection of meaningless procedures that must be learned by rote. Instead, the curriculum is *thinking* about mathematics. Mathematics education should involve helping students to think mathematically and to think about their mathematical thinking. This thinking curriculum requires a constant focus on (and assessment of) thinking, not on mathematics as a finished, static product. *Thinking assessment requires the thinking curriculum and the thinking curriculum requires thinking assessment.*

Notes

1. *Editors' note*: Some authors use the phrase *task-based interviews* to emphasize the fact that attention usually focuses on some mathematical problem or situation, in order to put the child's thinking into a more definite context.

2. *Editors' note*: It could be argued that the second version of the question also contains part of the answer, because the phrase "you add on a ten" may suggest the operation of addition to a child who did not think of it in response to "and I give you . . . ."

3. *Editors' note*: It is also apparent that Peggy uses the *sizes of the numbers themselves* as a guide and does not rely solely on the actual structure of the problem. Children commonly do this, but it is not always a good idea.

4. *Editors' note*: Of course, if this discussion took place without referring to a drawing of a number line, the child might not see why the "arithmetical" act of counting was being used in relation to the essential "geometrical" idea of "how close" the numbers are.

References

Artzt, A. F., & Armour-Thomas, E. (1990). *Development of a cognitive-metacognitive framework for protocol analysis of mathematical problem solving in small groups*. Paper presented at the annual meeting of the American Educational Research Association, Boston.

Budoff, M. (1987). Measures for assessing learning potential. In C. S. Lidz (Ed.), *Dynamic assessment: Foundations and fundamentals* (pp. 173–195). New York: Guilford Press.

Campione, J. C., Brown, A. L., Ferrara, R. A., & Bryant, N. R. (1984). The zone of proximal development: Implications for individual differences and learning. In B. Rogoff & J. V. Wertsch (Eds.), *Children's learning in the zone of proximal development*. New Directions for Child Development. San Francisco: Jossey-Bass.

Carlson, J. S., & Widaman, K. F. (1986). Eysenck on intelligence: A critical perspective. In S. Modgil & C. Modgil (Eds.), *Hans Eysenck: Consensus and controversy* (pp. 103-132). Philadelphia: Falmer Press.

Cocking, R. R., & Chipman, S. (1988). Conceptual issues related to mathematics achievement of language minority children. In R. Cocking & J. P. Mestre (Eds.), *Linguistic and cultural influences on learning mathematics*. Hillsdale, NJ: Lawrence Erlbaum.

Dionne, J. J., & Fitzback-Labrecque, M. (1989). The use of "mini-interviews" by "orthopädagogues": Three case studies. *Proceedings of the eleventh annual meeting of PME-NA*, New Brunswick, NJ.

Ericsson, K. A., & Simon, H. A. (1980). Verbal reports as data. *Psychological Review*, 87(3), 215-251.

Feuerstein, R. (1979). *The dynamic assessment of retarded performers: The learning potential assessment device, theory, instruments, and techniques*. Baltimore, MD: University Park Press.

Frederickson, J. R., & Collins, A. (1989). A systems approach to educational testing. *Educational Researcher*, 18, 27-32.

Ginsburg, H. P. (1990). T*he test of early mathematics ability: Assessment probes and instructional activities*. Austin, TX: Pro-Ed.

Ginsburg, H. P., & Baroody, A. J. (1990). *The test of early mathematics ability*, 2nd ed. Austin, TX: Pro-Ed.

Ginsburg, H. P., & Mathews, S. C. (1984). *Diagnostic Test of Arithmetic Strategies*. (DTAS). Austin, TX: Pro-Ed.

Ginsburg, H. P., & Opper, S. (1988). *Piaget's theory of intellectual development*, 3rd ed. Englewood Cliffs, NJ: Prentice-Hall.

Johnston, W. B., & Packer, A. E. (Eds.). (1987). *Workforce 2000: Work and workers for the twenty-first century*. Indianapolis, IN: Hudson Institute.

National Council of Teachers of Mathematics. (1989). *Curriculum and evaluation standards for school mathematics*. Reston, VA: Author.

National Research Council. (1989). *Everybody counts: A report to the nation on the future of mathematics education*. Washington, DC: National Academy Press.

Newell, A., & Simon, H. (1972). *Human problem solving*. Englewood Cliffs, NJ: Prentice-Hall.

Peck, D. M., Jencks, S. M., & Connell, M. L. (1989). Improving instruction through brief interviews. *Arithmetic Teacher*, 37, 15-17.

Piaget, J. (1929). *The child's conception of the world*. New York: Harcourt, Brace & World.

Polya, G. (1945). *How to solve it*. Garden City, NY: Doubleday.

Schoenfeld, A. H. (1983). Episodes and executive decisions in mathematical problem solving. In R. Lesh & M. Landau (Eds.), *Acquisition of mathematics concepts and processes* (pp. 345-395). New York: Academic Press.

Schoenfeld, A. H. (1985). *Mathematical problem solving*. Orlando, FL: Academic Press.

Silver Burdett & Ginn. (1991). *Mathematics: Exploring your world. Alternative assessment activities*. Morristown, NJ: Author.

Stigler, J. W. (1988). Research into practice: The use of verbal explanation in Japanese and American classrooms. *Arithmetic Teacher*, 36(2), 27-29.

Swanson, H. L. (1991). Issues and concerns in the assessment of learning disabilities. In H. L. Swanson (Ed.), *Handbook on the assessment of learning disabilities* (pp. 1–20). Austin, TX: Pro-Ed.

Vygotsky, L. S. (1978). *Mind in society: The development of higher psychological processes*, edited by M. Cole, V. John-Steiner, S. Scribner, & E. Souberman. Cambridge, MA: Harvard University Press.

CHAPTER THIRTEEN

Real Tasks and Real Assessment

JAN de LANGE

Abstract

Real school mathematics means also real assessment. Tests usually function in the margin of instruction but should be integrated into the instructional process. The tests offer results and feedback, not only for those students who can hit on clever strategies or proper ideas, but also for teachers who can get better information about the students' understanding. Examples from both standardized and nonstandardized tests from the Netherlands and the United States show that the case for authentic assessment is a complex one. Multiple-choice tests should be abandoned. Open questions are possible on many different levels, and students should be given ample freedom in constructing solutions. The Dutch exams show that standardized tests can offer much more than just repeated operationalization of the traditional basic skills. Higher order thinking skills can be part of standardized tests, although this test format has its restrictions and limitations. Further exploration shows that take-home tasks, two-stage tasks, and student-constructed tasks provide many opportunities for authentic testing. However, the construction and scoring of such tasks is not easy. Intersubjective scoring could replace traditional computer scoring.

Assessment of Teachers

During inservice teacher training courses in the early 1980s, the following excerpt from a college textbook on biology was given to university-educated upper secondary math teachers:

> . . . it might be interesting to estimate the number of offspring produced by one pair of rats under ideal conditions. The average number

---

All illustrations in this chapter are reproduced with permission of the Freudenthal Institute (OW & OC), Utrecht, The Netherlands.

of young produced at a birth is six; three out of those six are females. The period of gestation is twenty-one days; lactation also lasts twenty-one days. However, a female may already conceive again during lactation, she may even conceive again on the very day she has dropped her young. To simplify matters, let the number of days between one litter and the next be forty. If then the female drops six young on the first day of January, she will be able to produce another six forty days later. The females from the first litter will be able to produce offspring themselves after a hundred and twenty days. Assuming there will always be three females in every litter of six, the total number of rats will be 1808 by the next first of January, the original included. . . .

The question asked of the teachers was:

Is the conclusion that there will be 1,808 rats at the end of the year correct?

Only 20% of the teachers was able to solve this problem within half an hour. They explained, "We feel we have all the tools to solve the problem, but we are unable to use them." Here we saw excellent math teachers clearly embarrassed by the fact that although they were well educated in mathematics, were experienced teachers, and knew perfectly well how to design timed written tests and to prepare the students for those tests, they were nevertheless unable to "prove" whether or not the statement in the article was correct. The experience made the teachers more keenly aware of how limited and unrealistic our assessment procedures are. We prepare our students for a very narrow set of skills; as soon as we cross the border to real problem solving, more process-oriented skills, and unknown context areas, we are in deep trouble.

We have no overall knowledge about the ability of 16-year-old students to solve the problem, but the information from some of the experimental schools suggests that certain nonmathematically inclined students did very well on the problem. These students, it must be noted, were the first who were prepared by the new Mathematics A Curriculum, which tries to prepare students for using their mathematics in fields like social sciences, economics, and medical sciences (de Lange, 1987). Of course, the results also depend on the conditions. In the classroom, with a limited amount of time, students find it very difficult to solve or even to schematize the problem. But without a time limit—for instance, when the problem is given as homework—some fine results can be collected. These results indicate that such process-oriented activities are not well suited for testing by means of restricted-time written tests.

One girl came up with a surprisingly simple solution (Figure 13-1).

FIGURE 13-1

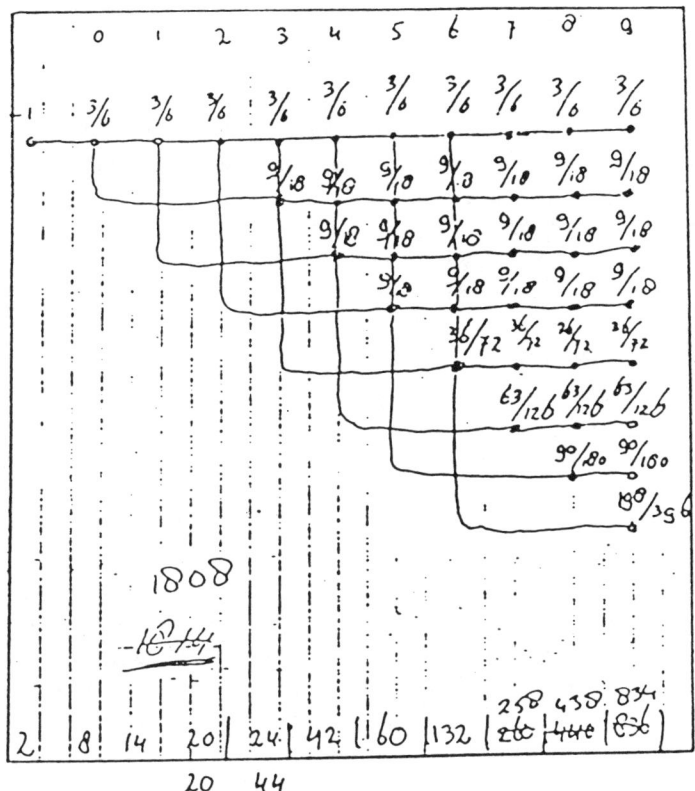

Another more schematized (instead of visual) solution, shown in Figure 13-2, came from a teacher. This solution suggests some kind of a formula, the need for which was strongly expressed by the teachers. They offered the following:

FIGURE 13-2

t	-1	0	1	2	3	4	5	6	7	8	9
N	2	6	6	6	1.2 6· 24	5.1 4 6· 44	1.10.6· 60	1.46· 132	1.86 6· 258	1.10 6· 438	1.27 8·6 834
T	2	8	14	20	44	86	146	278	536	974	1808

$$A_{n+3} = A_{n+2} + 3 \cdot A_n = 2$$
$$A_0 = 8$$
$$A_1 = 14$$

A completely different approach uses graphs and matrices (which are also included in the curriculum). The graph in Figure 13-3 represents in matrix form the growth of the rat population.

Another possibility is to look for the nature of the growth process. Comparing the number of rats period by period, we find that the growth factor in the long run is equal to 1.86. This leads to the formula:

$$A_n = 44 * 1.86^{n-3}$$

We leave it to the reader to integrate and generalize the different solutions, an activity that can be seen as the top level of mathematization.

Assessment

We began with experiments that eventually led to a new curriculum in the Netherlands for upper secondary students. In the process, we faced a serious problem: the restricted time available for written tests. We are not describing multiple-choice tests; these were never used in the first place because we judged that they were unfit for proper testing, especially for the higher levels. (As an example, consider the rat problem.) Even within the possibilities of the timed test, it was very difficult to design proper tests. When the design of tests was left to the teacher, the results were disappointing: Only 15% of the exercises really tested at a level other than the lowest one. So we started our developmental research to design new test formats. We followed these principles:

1. Tests should be an integrated part of the learning process, and therefore tests should improve learning.

FIGURE 13-3

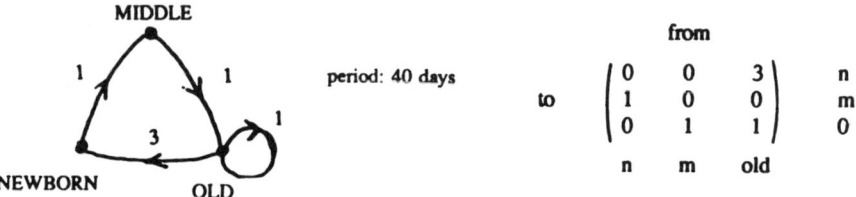

2. Tests should enable students to show what they know, rather than what they do not know. (We call this *positive testing*.)
3. Tests should operationalize all goals.
4. The quality of the test ought not to be dictated by its possibilities for objective scoring.
5. Tests should be practical in the sense that they should fit within the constraints of school practice.

Similar criteria were later also used in the test development of the "More project" for the elementary school level (the full name of the More project is "Research on the Effects of Traditional and Realistic Methods of Teaching on the Elementary Level"). We will first discuss this project in more detail, leaning heavily on earlier publications by Van den Heuvel-Panhuizen (1990) and Van den Heuvel-Panhuizen and Gravemeijer (1991).

The More project focuses on (among other things) the fact that the usual class-administered tests reveal only the bare results and tell nothing about the children's strategies. As a consequence of this lack of information on the children's strategies, wrong conclusions are likely to be inferred on the basis of children's performance. Another consequence is that too little information can be obtained about the progress of instruction; for instance, nothing is learned about the children's informal knowledge and problem-solving methods (Ginsburg, 1975). The last consequence is that the diagnosis of children's mathematical reasoning, when limited to answers, depends solely on the *results*—that is, does not reveal the children's difficulties and/or misconceptions.

As an alternative to the unsuccessful written tests that were developed, one nowadays pleads for individual observation and interviews as a mode to investigate children's knowledge and abilities. This, however, is only a partial solution. Among alternative modes of evaluation available to the teacher, written tests cannot be discarded. The following examples are taken from the More project for the third-grade students (8-year-olds). The tests are intended to be administered by the teacher to the whole class. Each student receives a booklet, and the teacher gives a brief oral explanation of each item.

Figures 13-4 through 13-6 provide sample items. The item in Figure 13-4, for example, shows two children comparing their height. It is related to "adding and subtracting." The obvious question is: How much is the difference? Figure 13-5, which relates to "ratio," asks: How many sweets are there in the bottom roll given that there are 18 sweets in the top roll? Measurement and geometry are the topics (see Figure 13-6) that are considered by posing the question: How many packets of chocolate are left after filling the box.

The realistic contexts in which these problems occur, contribute to the accessibility of the test problems while avoiding the obstructions caused by formal notation. Furthermore, the information provided about the children's

FIGURE 13-4

FIGURE 13-5

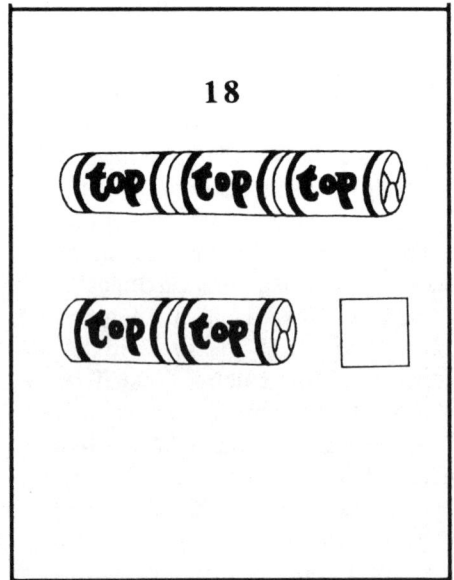

FIGURE 13-6

.... **pakken over**

informal knowledge and solutions helps the teacher to prepare the forthcoming instruction properly.

The next three examples, in Figures 13-7, 13-8, and 13-9, show that extensions in depth are possible. In Figure 13-7, the question is: How many packets of chocolate are needed for 81 children? First, the arithmetic operation is not given; moreover, even a correct calculation of 81/6 does not give the desired answer.

Figure 13-8 shows a more controversial example, by conventional test standards. The problem is open-ended and lacks explicit information. The question here is: Estimate the height of the neon letters on the building. Because the data needed for the calculation are lacking, the children are challenged to appeal to their knowledge of measures.

Another example of a higher order thinking problem is the item about the small train in Figure 13-9. A short ride takes 10 minutes to complete; the question is: How much time will it take to complete the long ride? Although the question is simple in itself, it gives no indication of the operation that must be carried out. On the contrary, a good analysis of the problem is required. How difficult this can be becomes very clear as we consider the introductory 'Rat' problem.

An advantage of test items like these is that they offer a variety of

270 Part Four • The Reality of Students

FIGURE 13-7

FIGURE 13-8

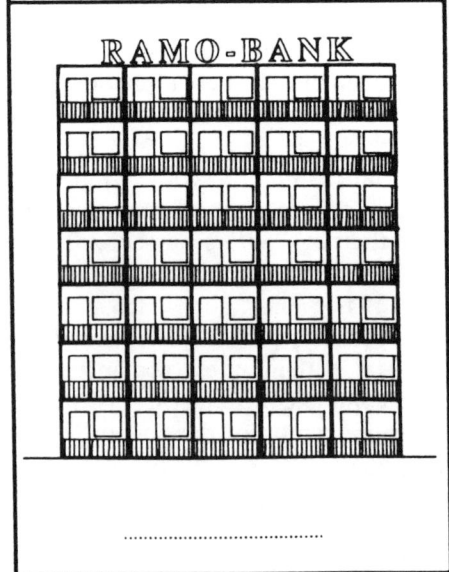

FIGURE 13-9

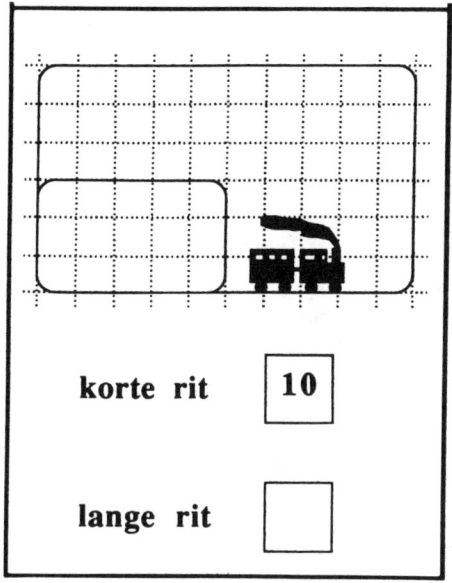

possible solutions at different levels. In this way it offers opportunities for children to show what they know rather than what they don't know. We call this approach to testing "positive testing."

All of the examples presented here come from a point of view that we refer to as "realistic mathematics education" (Treffers, 1986; Streefland, 1990; Gravemeijer, 1990; de Lange, 1987). One of the essential elements of this approach is that we offer students problems with more than one solution path. We abandon the idea associated with written tests in which there are unique answers for each item. The use of items that are sufficiently open-ended to allow several interpretations challenges students to think about alternative solutions. This not only encourages the children to show what they know, but also makes tests more informative in the sense that the solutions the children give are at their own level. For example, a great variety of answers is also possible on an item about a family in which the mother, father, and two children went to the circus and paid 50 guilders altogether. The question posed was to specify the admission charges for the tickets. Some children decided that all of the people would pay 10 guilders and the fourth person would pay 20 guilders. Others distinguished between admission costs for adults and for children and arrived at 13 and 12 guilders, or 15 and 10 guilders, respectively. Still others divided the total cost equally among the four persons to get 12.50, although decimal and other fractions had not yet been taught in school. Figures 13-10, 13-11, and 13-12 present some results by the students on the circus item.

National Standardized Tests

The More examples were illustrated for 8-year-old children. Now, turning to students 18 years of age, we look at some exercises from the new experimental final examinations that are in line with the philosophy of realistic mathematics education, or at least a step in that direction. In the examples, we will first discuss some exercises from a national final examination. These are interesting from various points of view. In the Netherlands there is a tradition that there are final, nationwide examinations at the end of four, five, or six years of secondary education. Roughly speaking, the six-year curriculum prepares the student for university, the five-year curriculum for higher vocational training, and the four-year course for the lower vocational level. Of course, many students just start working right after secondary education. Since 1985 we have had two new curricula for the six-year course. The experiments (carried out by OW & OC) that led to the new curricula clearly

FIGURE 13-10

Chapter Thirteen • Real Tasks and Real Assessment   273

FIGURE 13-11

FIGURE 13-12

showed, for the first time, the problems of achievement testing. The first nationwide examination took place in May 1987. Before that time, examinations were of the restricted-time written test kind (RTWT) with so-called open questions; they contained no multiple-choice items. The open questions, however, were in fact quite closed because both the answer (a number, a graph) and the solution left no degree of freedom whatsoever. Since 1987 modest progress has resulted in a somewhat more open final examination, still of the RTWT kind but also more problem- and text-oriented.

In the summer of 1990 two new curricula for the five-year stream were to be introduced; one (B) for the more mathematically gifted students and the other (A) for those who are going to use their mathematical skills in a nonmathematical profession or schooling. The first experimental nationwide examination took place in May 1989 (in this experimental setting, "nationwide" meant only three schools) and showed that some progress was indeed being made. The exercises are designed according to the following criteria:

1. They are more open in the broader sense.
2. They try to assess also some higher order thinking skills.
3. They offer students more possibilities to solve the problem at their own level (differentiation).

A typical final examination in mathematics (A or B) takes three hours to complete. The exam consists of about five big problems with approximately 20 questions on some six pages of text. Because of the space limitations of this chapter, we will show only some of the shorter problems.

From the A Exam

If no fish are caught, the number of fish will increase in the coming years. The graph in Figure 13-13 shows a model of the growth of the number of fish. Draw an increase diagram with intervals of one year, to start with the interval 1–2.

The fish farmer will wait some years before he catches or harvests any fish. After the first catch he wants to catch the same amount of fish every year as he did the first year, and he wants to catch as much as possible. After every catch the number of fish increases again according to the graph. The exercise is as follows:

How would you advise the fish farmer about:

1. The number of years he has to wait after planting the fish, and
2. The amount of fish he will catch every year?

Give convincing arguments.

FIGURE 13-13

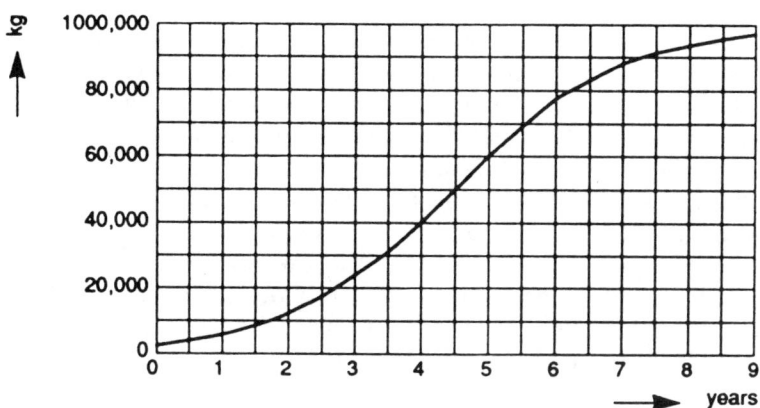

When considering this exercise, one should know that the curriculum does not cover the subject "differentiation of functions" but includes the changes of real phenomena in a discrete way (which provides excellent preparation for the students who will study calculus). Instead of the graph of the derivative of a function, the students are accustomed to the discrete apparatus called an increase diagram. So the first question is very straightforward and operationalizes only the lowest level. The other question was new and was long awaited, in our opinion. Communicating mathematics, drawing conclusions, and finding convincing arguments are activities that are too seldom found in mathematics tests and examinations. Many teachers were surprised and did not know what to think of this new development, although they were prepared. The experiments gave proper indications of the new approach. Students, as usual, seemed less surprised, although they gave a wide variety of answers:

> I would wait for four years and then catch 20,000 kg per year. You can't lose that way, man.

> If you wait till the end of the fifth year, then you have a big harvest every year: 20,000 kg of fish; that's certainly not peanuts. If you can't wait that long, and start to catch one year earlier, you can catch only 17,000 kg, and if you wait too long (one year) you can only catch 18,000 kg of fish. So you have the best results after waiting for five years. Be patient, wait those years. You won't regret it.

From the B Exam

The students who take the B exam are preparing for a higher technical vocational school; in general, they will need quite a bit of more formal and abstract math for their job or schooling. Besides that, some three-dimensional insight won't do them any harm. The next exercise, indicated in Figure 13-14, gives some insight into what we expect from our students in that respect.

The large tower of the church in the picture is the subject of more detailed study. The floor of this tower is a square, 6 by 6 meters. The roof is formed by four equally sized rhombuses. The lowest vertices of these roof parts are at a height of 18 meters above the ground. The top is at a height of 26 meters. The other four vertices of the roof are at a height of 22 meters, each on the axis of symmetry of the sidewalls.

On the worksheet you see the beginnings of a drawing of the tower, in the so-called engineers projection. Finish the drawing of the tower.

The upper gaps in the walls of the tower are the reverberation holes. Behind those gaps hangs the bell that is rung every half hour. The quality of the sound depends on the shape and volume of the bell room. The floor of this room is at a height of 12 meters above the ground. The ceiling can be constructed at a height of 20, 22, or 24 meters.

Draw to scale 1:100 in one figure the shape of the three possible ceilings. The ceiling is placed at a height of 22 meters.

Compute the volume of the bell room.

FIGURE 13-14

Chapter Thirteen • Real Tasks and Real Assessment   277

This exercise is less open than the previous one but certainly involves a reasonable level of three-dimensional geometry that is somewhat different from traditional geometry. The difference is that the students need to apply their knowledge in a real-world problem. The passage from mathematics to its applications all too often involves crossing a gap for which too few bridges have been constructed. The tower is real, the dimensions are real, the problems are more or less real, and the mathematics may be considered from a variety of levels of sophistication.

Figures 13-15 and 13-16 show the worksheet (at left) and two different solutions offered by students.

If the trend toward more open examinations continues, it will definitely affect the teaching of mathematics. As in many other countries, the Dutch teacher (or school) is judged by how well the students perform on their final exams. This leads to test-oriented learning and teaching. If the test is made according to our principles, however, this disadvantage (test-oriented teaching) will turn into an advantage. The problem then shifts to the producers of tests. In the Netherlands tests are produced by a government-funded independent institute for achievement testing (CITO), which in turns tries to cooperate as much as possible with others involved in mathematics education and, not least, with OW & OC.

One key factor that remains is the teacher, who must accept wholeheartedly the new emphasis on open-ended, complex problems. Such

FIGURE 13-15

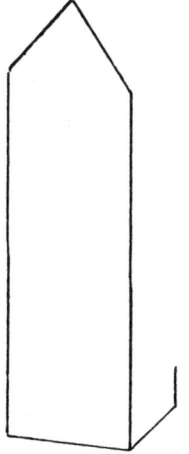
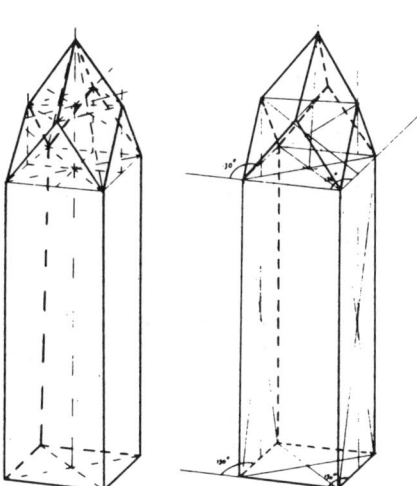

FIGURE 13-16

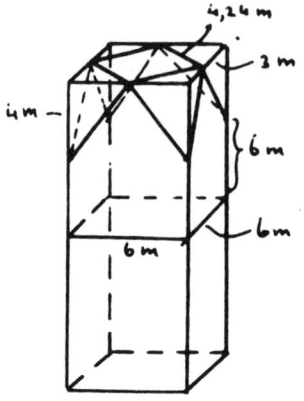

Floor area: $\frac{1}{2} \cdot 3 \cdot 3 = 4.5$
Vol. $= \frac{1}{3} \cdot A \cdot h$.
$\frac{1}{3} \times 4.5 \times 4 = 6 \text{ m}^3$
$5r : 4 \cdot 6 = 24 \text{ m}^3$
Vol block: $6 \cdot 6 \cdot 10 = 360$
Total: $360 - 24 = 336 \text{ m}^3$

problems make teaching more difficult and complex as well. The teacher may view this as a loss of some authority because students will come up with "smart" solutions. The teacher also will have to interact more with students to discuss different solutions. Thus, even if the test producers succeed in creating better achievement tests (at a national level), the teacher remains the crucial factor. Teachers deserve a lot of attention and help in designing school tests. Because there are fewer restrictions, the teacher has a wider variety of possibilities for designing and administering tests. This variety makes the test problems more exciting and rewarding but, again, poses more problems for the teacher.

Now, let us shift our attention to school tests.

School Tests

For testing at the lowest level, the RTWT remains a very good tool (excluding multiple-choice questions). Even with those unfavorable conditions, students can perform on higher levels when given the proper tests. The following example is taken from a 50-minute school test that consisted of three problems. The students were taking the A curriculum and were 16 years old.

Figure 13-17 shows a crossroads in Geldrop, The Netherlands, near the Great Church. The traffic lights have been regulated in order to avoid rush hour traffic jams. A count showed how many vehicles per hour had to pass the crossroads during rush hour (Figure 13-18). In Figure 13-19 the matrices G1, G2, G3, and G4 show which directions have a green light and for how long. Two-thirds (2/3) means that traffic can ride through a green light for a period of 2/3 minute. The questions are:

FIGURE 13-17

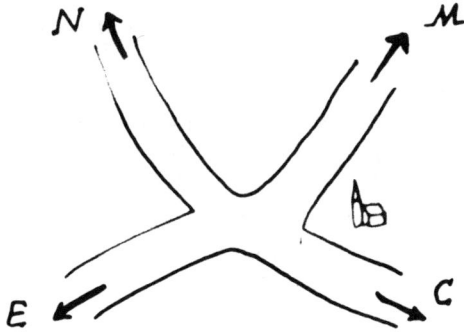

FIGURE 13-18

$$A: \quad \text{from} \quad \begin{matrix} M \\ N \\ E \\ C \end{matrix} \begin{pmatrix} 0 & 40 & 200 & 30 \\ 30 & 0 & 80 & 50 \\ 210 & 60 & 0 & 60 \\ 30 & 40 & 80 & 0 \end{pmatrix}$$
$$\quad\quad\quad\quad\quad M \quad N \quad E \quad C$$

FIGURE 13-19

$$G_1 = \begin{matrix} & M & N & E & C \\ M & \\ N & \\ E & \\ C & \end{matrix} \begin{bmatrix} 0 & \frac{2}{3} & \frac{2}{3} & 0 \\ 0 & 0 & 0 & 0 \\ \frac{2}{3} & 0 & 0 & \frac{2}{3} \\ 0 & 0 & 0 & 0 \end{bmatrix} \quad G_3 = \begin{matrix} & M & N & E & C \\ M & \\ N & \\ E & \\ C & \end{matrix} \begin{bmatrix} 0 & 0 & 0 & 0 \\ 0 & 0 & \frac{1}{2} & \frac{1}{2} \\ 0 & 0 & 0 & 0 \\ \frac{1}{2} & \frac{1}{2} & 0 & 0 \end{bmatrix}$$

$$G_2 = \begin{matrix} & M & N & E & C \\ M & \\ N & \\ E & \\ C & \end{matrix} \begin{bmatrix} 0 & 0 & 0 & \frac{1}{3} \\ 0 & 0 & 0 & 0 \\ 0 & \frac{1}{3} & 0 & 0 \\ 0 & 0 & 0 & 0 \end{bmatrix} \quad G_4 = \begin{matrix} & M & N & E & C \\ M & \\ N & \\ E & \\ C & \end{matrix} \begin{bmatrix} 0 & 0 & 0 & 0 \\ \frac{1}{2} & 0 & 0 & 0 \\ 0 & 0 & 0 & 0 \\ 0 & 0 & \frac{1}{2} & 0 \end{bmatrix}$$

1. How many cars come from the direction of Eindhoven during that one hour? How many travel toward the city center?
2. How much time is needed to have all lights turn green exactly once?
3. Determine $G = G1 + G2 + G3 + G4$ and thereafter $T = 30G$. What do the elements of T signify?
4. Ten cars per minute can pass through the green light. Show in a matrix the maximum number of cars that can pass in each direction in one hour.
5. Compare this matrix to matrix A. Are the traffic lights regulated accurately? If not, can you make another matrix G in which traffic can pass more smoothly?

Of course, this exercise bears all the marks of a RTWT. It is relatively closed and poses a number of short questions in order to guide the students. It would be interesting to find out what would have happened if we had posed only the last question.

The Milwaukee Project

A more recent project (1989–1990) involved the possibilities of "real" assessment in a school situation at a suburban Milwaukee school. The mathematical subject was data visualization. The students were ninth-graders (13 to 18 years old). A student text was developed by our OW & OC institute with the help of the Whitnall High School in Greenfield, Wisconsin. During the five-week course we carried out four different tests (de Lange, Van Reeuwijk, Burrill, & Romberg, 1991), as follows:

1. A restricted-time written test to be taken after the first two chapters of the booklet were treated. The test contained straightforward closed questions as well as some more open questions.
2. A restricted-time written test to be taken after the whole booklet has been done. Most of the questions were closed, but some were more open questions.
3. An extended-response essay task, after chapters 1 through 2 were treated. In our case the task was to rewrite a given story with proper use of data visualization.
4. The following test: Let the students construct a test. Tell them, after the two first lessons, to look for materials that are to be used in a test that each student is to construct for fellow students. The test should be completed some three weeks after completing the course. We used a somewhat different classification than the traditional ones. Closed questions in the usual definition were most often multiple-choice tests. We were not interested in this test format, for the background of these

tests was merely one of economic and financial reasons, and not content-related. So we will neither discuss nor deal with test questions of the multiple-choice format.

A closed question, in our definition, gives the student no freedom. Examples are:

1. Solve: $x^2 + 3x + 2 = 0$.
2. Draw the graph of: $y = x + 3$.
3. What is the average of: 4, 5, 6, 7, 3, 8.

An open question, in contrast, gives the student somewhat more freedom in solving the problem; sometimes there is more than one correct answer. The reasoning behind the solution can be more important than the solution itself. We will see examples of open questions in our tests. There seems to be some loose relationship between the use of closed versus open questions and the level of activities operationalized in the questions. In other words, it seems harder to test higher order thinking skills using closed questions than it is with open questions.

This is especially true if we want to test processes like mathematization. Process-oriented skills testing seems almost impossible with closed-question tasks. Open questions and, even more, essay questions offer much more opportunity in this respect.

There are very few tools we can use to test reflection. For the "free production" test, students are asked to construct a test for their fellow students; this encourages them to reflect on their own learning process. We also encourage them to formulate questions in a precise mathematical way. We give them an opportunity to show their creativity, even more than with essay questions. On the other hand, these tests can give the teacher valuable information on the way the students perceived the curriculum. They give the teacher important feedback about the teaching–learning process. In this way the "free production test" is an important tool if we see as a purpose of testing the improvement of learning (Streefland, 1987; Van der Brink, 1987; and more specific for the More project, Van den Heuvel-Panhuizen, 1989).

Certainly, no one-to-one relation exists between our proposed hierarchy—closed questions, open questions, essay questions, free production tests—and cognitive levels, but in general one can state that it is easier to test higher order thinking skills with more open questions.

We will now discuss the four different tests that were carried out during the experiment. We gave the following four graphs indicated in Figure 13-20 to the students. After some straightforward questions regarding basic skills, we presented the last question:

FIGURE 13-20

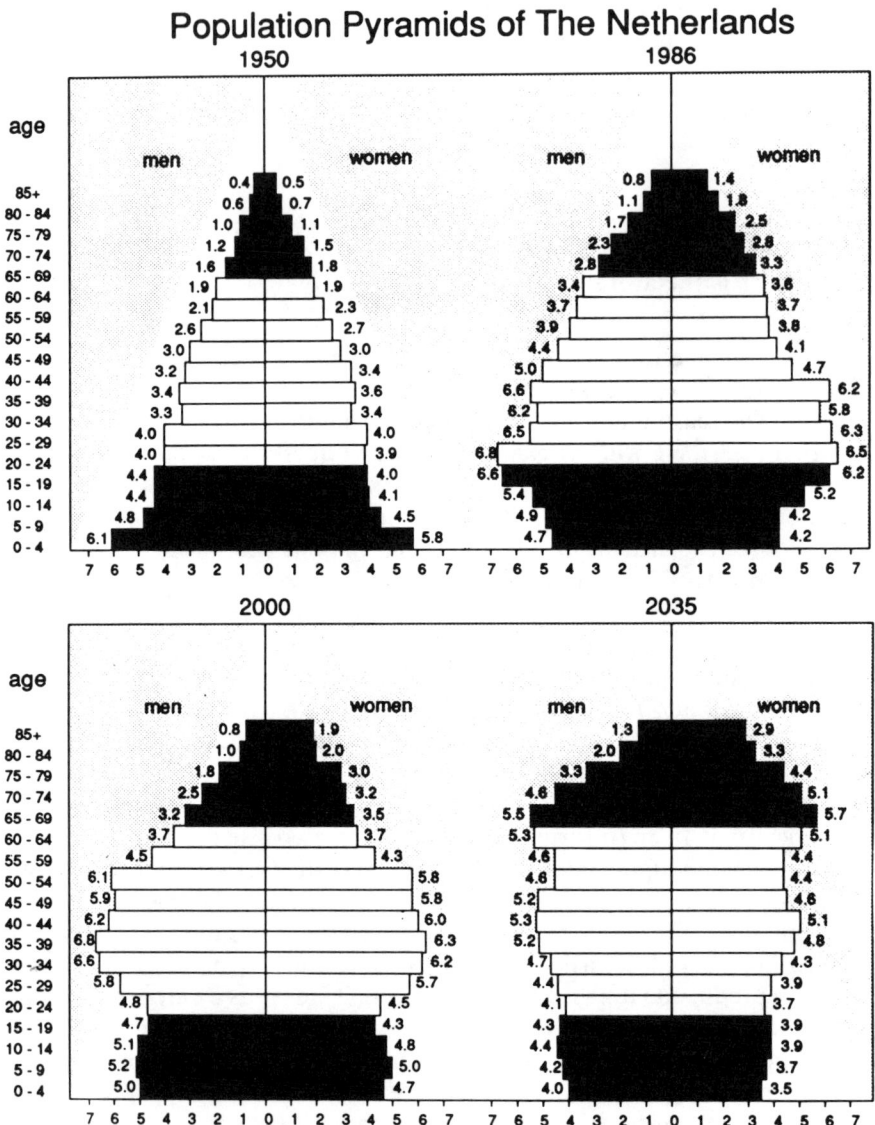

Numbers are in hundred thousands (100,000's)

Choose a graphical way to show clearly the problem of the aging population in the Netherlands.

This question is not simple to answer (as part of a timed school test), for the student must succeed in doing the following:

FIGURE 13-21

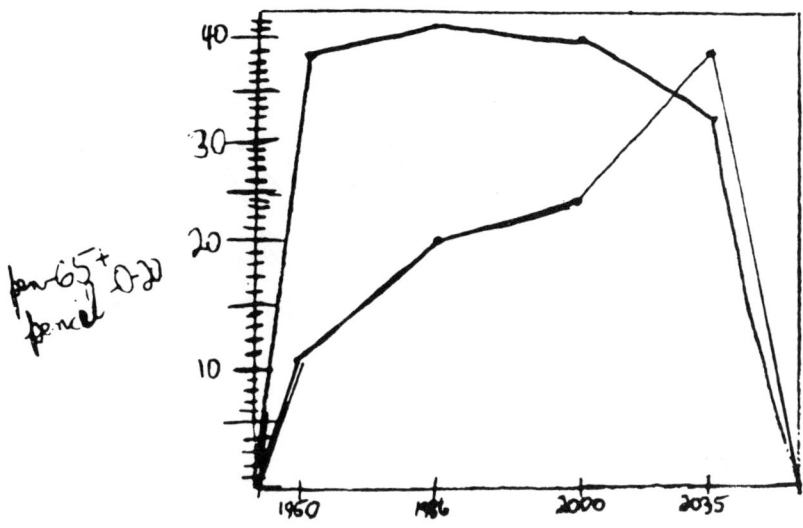

Understand the problem.
Compare the four (eight) graphs.
Decide how to visualize.
Decide how to divide men from women and scale the graph(s).
Draw the graph(s).

Some solutions that were presented are included in Figures 13-21, 13-22, and 13-23. Although the graph in Figure 13-21 is partially correct, the student has a "horizontal axis fixation": He wants to start at the origin and to finish on the horizontal axis as well. The graph in Figure 13-22 not only shows the aging problem but also gives details on both the female and the male perspective. A rather surprising bar graph is shown in Figure 13-23.

FIGURE 13-22

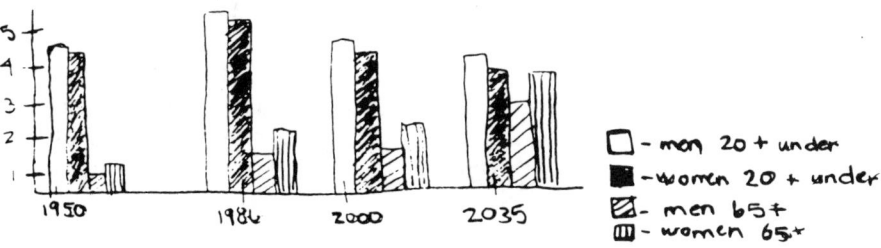

FIGURE 13-23

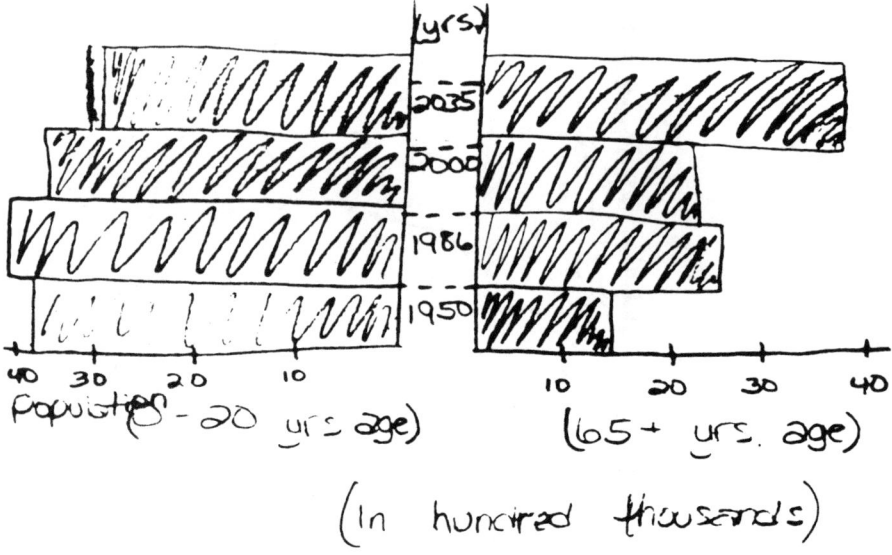

What was unusual about this question for both teachers and students? The answer was not a number or a simple graph. It was necessary to read a lot, look at four different graphs, make decisions, write sentences, think (there are no similar exercises in the book, so the trigger mechanism is lacking), and come up with a story or graph. There was no single correct answer.

The second test considered an article about mass migration in Indonesia. The students were asked to rewrite the article at home making proper use of data visualization. Two examples (Figures 13-24 and 13-25) show the uneven distribution of the population over the islands as carried out by the students.

The third test mainly operationalized the lower goals, although there were again some questions that were too difficult for those at the lower ability and skill levels. The students performed well on this test.

The fourth and final test was something completely different. We asked the students to design a test for their fellow students. Here are some of the problems that students made up:

1. If a hand-held calculator cost $30.00 in 1970, and a similar calculator cost $15.00 in 1980, how much would you expect it to cost in 1985?
2. What kind of graph is shown (in Figure 13-26)? What information is involved?

FIGURE 13-24

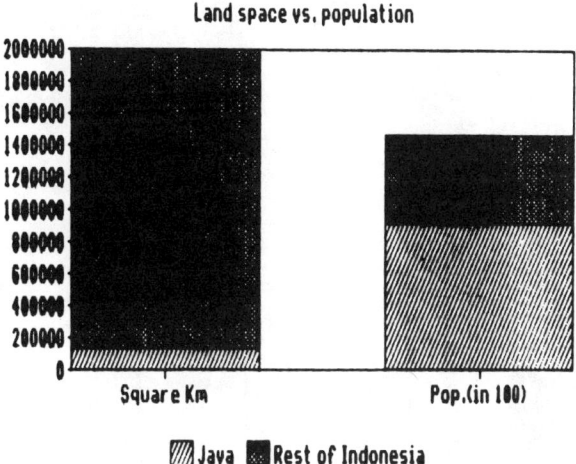

The graph shown here is a bar graph involving information about how much money is being spent on military supplies compared to the country's GNP.

[Answers by other students are shown in italics. All questions and answers refer to Figure 13-26.]

FIGURE 13-25

FIGURE 13-26

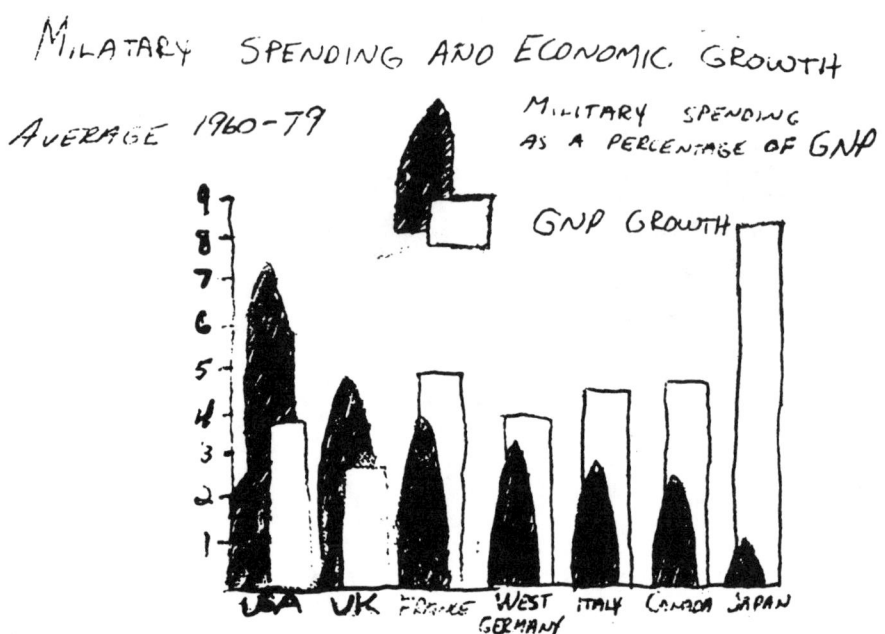

Is this information easy to read and understandable? Why or why not?
No, this information is not easy to read because the numbers on the left have no meaning. One cannot tell if it means millions, billions, etc.

Could this information be represented better? Explain.
No, because a line graph would not work, a box-plot, pie graph, stem-leaf, etc.

Is this graph accurate? Explain your answer.
No, because this graph is based on an average between 1960 and 1979.

Conclusions

It will be clear from our examples that tests can be an integral part of the instructional process. They can offer exercises, not only for the students, who can hit upon clever strategies, but also for teachers, who can gain a better understanding of what students are able to do and can be put on the track of richer didactics. This could reverse the earlier roles played by tests. The examples from the Dutch national standardized examination show that timed written tests can offer more than just operationalizing the old basic skills. The road toward real assessment (or authentic assessment) and real tasks will be long and hard, but at least a start has been made.

References

de Lange, J. (1987). *Mathematics, insight and meaning.* Utrecht: OW & OC.
de Lange, J., Van Reeuwijk, M., Burrill, G., & Romberg, T. (1991). *The learning and testing of mathematics in a context.* Forthcoming.
Ginsburg, H. (1975). Young children's informal knowledge of mathematics. *Journal of Children's Mathematical Behaviour, 1,* 63-156.
Gravemeijer, K. (1990). Context problems and realistic mathematics instruction. In K. Gravemeijer, M. Van den Heuvel, & L. Streefland (Eds.), *Contexts, free productions, tests and geometry in realistic mathematics education.* Utrecht: OW & OC.
Streefland, L. (1987). *Free productions of fractions monographs. Proceedings PME 11.* Montreal.
Streefland, L. (1990). Realistic mathematics education: What does it mean? In K. Gravemeijer, M. Van den Heuvel, & L. Streefland (Eds.), *Contexts, free production, tests and geometry in realistic mathematics education.* Utrecht: OW & OC.
Treffers, A. (1986). *Three dimensions.* Dordrecht: Reidel.
Van der Brink, J. (1987). Children as arithmetic book authors. FLM, 7(2), 44-48.
Van den Heuvel-Panhuizen, M. (1989). Realistic arithmetic/mathematics instruction and tests. In K. Gravemeijer, M. Van den Heuvel, & L. Streefland (Eds.), *Contexts, free productions, tests and geometry in realistic mathematics education.* Utrecht: OW & OC.
Van den Heuvel-Panhuizen, M., & Gravemeijer, K. (1991). Tests aren't at all bad. *Arithmetic Teacher.*

Index

A and B final examinations, 274–278
Activity-based problem-solving episodes, 173
Additive inverse, finding the, 55
Adler, Mortimer, 44
African-American students, 23–24
Agriculture, problems about debt and loss in, 66–67
Algebraic notation, confusions in, 55–57, 215
Algorithms, students', 18
American Association for the Advancement of Science, 219
Approaches to early childhood learning, 14
Arithmetic skill, decrease in value of, 81
Assessment
 of children's knowledge, 109–112
 in the classroom, 253–256
 and Cognitively Guided Instruction, 110–112
 formal and informal, 110, 111
 and limits of restricted time for written tests, 266
 of mathematical thinking and learning potential, 237–260
 of mental processes, 241–253
 methods of, effectiveness of various, 254
 needed changes in, 229, 238–239
 new approaches to, 238–239
 as part of instructional day, 112
 real, 263–287
 of teachers, 263–266
 systems of based on rote model, 3, 239
Attributes, 52–54

Basic mathematics skills, need for, 219
 Basic tasks, failure to establish, 20
Behavior, problem-solving, 103
Beliefs about mathematical activity, 120, 126–127, 218
Bell, Alan, 225

Calculator technology
 effective use of, 5
 new skills required by, 219
Calculus, concept of limits in, 217
Cancellation, 53
Caring
 as ethical framework for constructivist pedagogy, 36, 43
 and pedagogy, 43–49
 in relation of student and teacher, 42, 47
Change
 commitment to positive, 2
 computers and, 80
 in curriculum, 221–222
 in educational system, 1
 in pedagogy, 221–222
 in societal expectations, 230
Children's
 strategies used by, 267
 thinking of, as basis for instructional decisions, 113–114, 239
 thought processes of, in mathematics instruction, 97
 understanding of minus sign by, 62
Civil War and assimilation paradigms, 28
Classroom, assessment in, 253–256
Clinical interview, the, 241–248
Coercion vs. caring, 45
Cognitive apprenticeship, 42
Cognitively Guided Instruction (CGI), 93–115
Cognitively oriented approaches, 3
Cognitive phases, 223–224
Cognitivist view of learning, 220, 224
Communication
 mathematical, 123, 141, 160
 during mathematical problem solving, 138
 as integral part of learning, 137
Compartmentalization, 4, 220–221
Complacency, national, 2
Computation without meaning, 216
Computer graphics
 as medium for teaching, 85
 misconceptions formed by, 222
Computer technology
 effective use of, 5, 79–92
 new skills required by, 219
Computing as a paradigm of science, 80
Concatenating, 59
Concepts learned outside of school, 68
Conceptual change classroom, 224–225
Conceptual understanding, need for, 3
Concurrent verbalization, 248
Conflict/growth, cognitives, 222
Conflict teaching, 225–227
Connections, forging, 46
Conservation
 steps in development of, 223
 of operation, 216
Constructivism
 and caring, 35–49
 emphasis on students' communication in, 136
 ethic of, 35–36
 roots of, 136–137
 and the teaching-learning process, 139
Constructivist pedagogy, 40–43
Constructivist researchers
 ethical obligations of, 39
 in mathematics education, 139, 219–220
Constructivist teaching, example of, 38–39
Continuous variable, 59
Convincing arguments, as part of testing, 275

Index 289

Crisis in mathematics education, 1–7
Cube example, 154–162, 167–171
Cuisenaire rods, 22, 124, 186, 199–209
Cultural barriers in mathematics profession, 4
Curricula, two kinds of mathematics, 274
Curriculum
  development of, and teacher education, 95
  development of traditional, 3
  limits of explicit, 16–17, 26
  mathematics and the, 105–106
  modernizing the, 12, 221–222
  toward the thinking, 257–259
  that undermines student understanding, 220
  zero-based mathematics, 79, 82–85
Curriculum and Evaluation Standards for School Mathematics, 106–107, 219

Data
  analysis of, need to teach in elementary school, 86
  visualization of, 284, 285
Descartes, Réné, 46, 53
Development of mathematical knowledge, 173–210
Dewey, John, 36–37
Diagnostic Test of Arithmetic Strategies, 248
Direct instruction teacher, 98
Disagreements
  identifying and coding, 144–148
  scores for, 148, 150–151
  students', 135–171
Discovery learning, 87
Discrepancies, experience of, 223
Division, misconceptions about, 215
Drill
  and practice, 3, 89
  vs. problem solving, 108
Driver, R., 222, 225

Education crisis, mathematics, 1–7
Educational goals, low, 1
Effros, Edward, 88–89, 219
"Emergent literacy" approach, 14
Environment, nonthreatening, 225, 229
Equals sign, understanding of, 215, 240–241
Errors, occurrence of, 18
Escalante, Jaime, 44–45, 48
Estimation abilities, 90
Ethical obligation in pedagogy, 37–49
Ethic of shared learning, 254
Evaluation, 49
Explanations vs. justifications, 121–123
Explicit curriculum, limits of, 16–17, 26
Exploration, benefits of free, 209

Feuerstein, R., 251
Fischbein's didactical implications, 228
Fish farmer problem, 274
Flavell, J., 222, 223–224, 225, 226, 227
"Focusing on skills" approach, 14
Fourth National Assessment of Educational Progress, 219

Gains and losses, 51–53
Geometric figures, stereotypic images of, 216
Geometry, Euclidean, 46
Goals
  academic vs. relational, 47
  low educational, 1
  in mathematics education, 218
  open-ended, in flexible interviewing, 242
  and test scores, 3
Group activities and autonomous thinking, 139

Group dynamics and problem solving, 151
Group interaction and problem solving, 154
Guess-and-check strategy, 183

Helplessness, learned, 220
High School Proficiency Test (HSPT), 10
Human Math Machine, 107

Ideas about changing ideas, 226
Information, pace of, 222
Information-processing activities, 27
Innovation in teaching, opportunity for, 86
Input data, mapping of, 21, 27
Inquiry vs. school mathematics, 123, 130
Insights, sharing mathematical, 159
Instructional behavior of Ms. G., 101–103
Interviews
  clinical, 241–248
  as methods of assessment, 239–246, 254–255
  mini-, using the, 254–255
  and sociometric questions, 166–167
  structured, 244–246
  task-based, 174
Interviewers, students acting as, 255
Interviewing, flexible, 242–244
Interviews with Brian in grade 6, 185–204
Intuition, importance of valuing, 220, 228
Intuition in Science and Mathematics, 228

Journal of Children's Mathematical Behavior, 214
Judgmental listening vs. passive acceptance, 16
Justifications vs. explanations, 121–123

Knowledge
  theories of, 36
  two kinds of, 27–28
Kuhn, Thomas, 82

Lakoff, George, 27
Landis study, the, 175–176
Law of the Ineffective Middle, 20
Learners' cognitions in mathematics, 94
Learning environments, 141
Learning experiences
  designing, 222
  kinds of, 14
Learning mathematics
  complexity of, 14–17, 18–19, 210
  perspectives from everyday life, 61–77
  with understanding, 120
Learning potential
  limited, 253
  techniques for assessing, 239, 250–253, 256
Lesson schemes, Driver's, 224–225
Lewis, C. I., 36
Lomotey, Kofi, 11, 23

Madison Project, 18
Manipulative materials, use of, 119, 124, 129, 242
Mapping
  conceptual, 58
  of input data into schemas/paradigms, 21, 27, 182, 191
Materials
  instructional, 229
  for small-group problem solving, 143
Mathematical activity
  beliefs about, 120
  Brian's, 173
  meaningful, 130
Mathematical community, mores of, 48
Mathematical knowledge
  opportunities to construct, 123

Mathematical knowledge *(Continued)*
  testing for, 246
  wider need for, 81
Mathematical learning
  communication in, 137
  student's, 5
Mathematical models, computing and, 80
Mathematical power, acquiring, 17, 25, 220
Mathematical problem solving and group interaction, 154
Mathematical profession, cultural barriers in, 4
Mathematical representations of problems, 13
Mathematical Science Education Board, 85, 219
Mathematical thinking
  assessment and, 260
  complexity of, 210
  inducing, 41
  learning to use children's, 93-115, 240
  student's, 175, 243, 245
  teachers' understanding of students', 14, 240
Mathematical truths, 128-132
Mathematical understanding, teaching for, 119
Mathematicians
  reluctance of to adopt technology, 80
  view of mathematics of, 57
Mathematics
  achievement in, 3, 238
  beliefs about, 119
  changes in due to technology, 80
  children's thinking about, 93-115
  communication of during testing, 275
  compared to foreign language, 61-62
  complexities of doing, 14, 18-19, 210
  and the curriculum, 105-106
  curriculum in, zero-based, 79, 82-85
  discrete, and children's thinking, 59
  educator's view of how to help students with, 57
  Greek, 55
  how students think about, 17
  modernizing curricula of, 12
  preparation in for other fields, 264
  as procedural instructions, 126
  real vs. school, 2, 40-43, 46-47, 214
  and the reality of the student, 213-231
  school, background of, 11-13
  school vs. inquiry, 123, 130
  situated, 51, 54-55
  students', 214, 218, 243, 245
  as taught, 106-107
  teaching of, decline in quality of, 83
  as a theoretical discipline, 80
  as thinking, analyzing, and planning, 210, 238
  thinking about, 237
Mathematics education
  assessment in, 237-260
  constructivist approaches to, 87
  crisis in, 1-7, 82
  evolutionary development of, 82
  relevance of technology in, 81-82
  research in, 5, 87-88
  rigid specification in, 12
  solving real problems of, 6
  technological impact on, 79-92
Mathematics teachers
  professional status of, 5, 97
  responsibilities of, 45
  technology and, 86-87
Meaning and symbols, 54-55, 220
Mediated learning, 251
Mental calculation, role of, 90
Mental processes, assessing, 241-253
Metacognition, 225, 257-258

Metaphors, basic, 27
Methodology, research, 26-29
Milwaukee project, 278-286
Minority students, programs for, 23-26
Minus sign
  and magnitude, 63
  meanings for, 55-57, 61-64
  and the operation of subtraction, 63
  representing inversion, 64
Misconceptions
  development of, 222
  difficulty in uncovering, 229-230, 267
Misunderstanding, problems caused by, 19
Monotonicity, 58
More project, 267
Morrow, Lesley, 14-16
Motivation, students', 40, 46
Motivational displacement, 43
Multiplication, misconceptions about, 215

National Council of Teachers of Mathematics (NCTM), 2, 15, 40, 88, 89, 107, 120, 238
National Research Council (NRC), 85, 88, 120
Negative numbers
  acceptance of, 53
  empirical studies involving, 64-74
  historical understanding of, 62
  notation for, 56
  reality of, 51-60
  studies involving, 64-74
Negative sign, meanings for, 55-57, 61-64
Neill, A. S., 44
New Jersey Business and Industry Science Education Consortium (BISEC)
Notation, algebraic, 55-57
Notational confusion, 75-76
Noun vs. verb knowledge, 21
Numbers
  associated with magnitudes, 62
  as manipulable arithmetical objects, 123
  signed, 65
Numerical relationships
  construction of, 120, 127
  recognition of, 253

Observation of students at work, 17-23
Open-ended complex problems, 269, 277, 281
Open examinations, trend toward, 277
Operations
  binary, 55-57, 58
  with fractions and decimals, 215
  of subtraction, meanings for, 65

Page, David, 56
Paradigms, assimilation, 27
Paradigm shifts in intellectual disciplines, 82
Pattern Blocks, 177, 183-185, 186-194
Patterns
  in Brian's approach to building solutions, 209
  as characteristic of development, 223
  habit of looking for, 13
  learning when to make use of, 15
  and relationships in mathematics, 141
Pedagogical assumptions, underlying, 13
Pedagogical devices, 59-60
Pedagogical methods, modernizing,d 12
Pedagogical strategies, 228
Pedagogy
  changes in, 86, 87, 221-222
  constructivist, 40-43
  ethical obligation in, 37, 38-39
  identifying appropriate, 85

software as part of, 91
Peek, B., 25
Piaget, J., 27, 38, 57, 64, 136, 139, 222, 239-240, 241-242, 248
Pierce, C. S., 35, 36
Place value
  children's thinking about, 94
  interpretation of two-digit numbers, 123
  numeration, 119, 130
Political education, 25
Popper, Karl, 36
Population pyramids of the Netherlands test, 282-284
Positive testing, 267, 271
Probability
  need to teach in elementary school, 86
  and Pascal, 46-47
Probes, organized, 246-247
Problems, habit of analyzing, 18
Problem solving
  age effect vs. school effect in, 66
  Brian's, 174
  with concepts learned outside of school, 68
  difficulties of inversion in, 65
  effect of schooling on, 66
  mathematical, 138
  from procedures to, 107-108
  real, lack of preparation for, 264
  research on communication during, 137-138
  small-group, 119, 120-123
  students' disagreement during small-group, 134-171
  testing for knowledge about, 267
  written vs. oral representation in, 66
Problem-solving ability, improved, 96
Problem-solving activities, 204
Problem-solving behavior, asking children to engage in, 103, 129
Problem-solving opportunities, learning extracted from, 141
Problem-solving process
  assistance in, 251
  reflection on, 257
Problem-solving strategies
  children's, 95, 183, 247, 248
  sharing of, 141
Problem type analysis, 94-95, 269
Procedural instructions, 123-128
Procedural requests, 153
Process-oriented skills
  lack of preparation for, 264
  testing, 281
Professionals, teachers as, 3-5
Profit and loss, understanding of, 65

Questioning one's solution, 176-177

Rational numbers, children's thinking about, 94
Real mathematics
  definition of, 41, 214
  vs. students' mathematics, 214-215
Real problem-solving, 264
Realistic mathematics education, 271, 272
Relationships
  constructing mathematical, 120, 127, 253
  one-to-one, 254
Reliance on outside authority, 220
Reports, retrospective, 249-250
Representation
  Brian's, 173-210
  building a, 179
Representations

of fractions, 204
importance of, 19
introducing new, 62
observable kinds of, 175-176
relating one to another, 178
symbolic, 208
Research
  in mathematics education, 5-6, 87-88
  methodology of, 26-29
  and practice, interrelation of, 88
  on student communication during problem solving, 137-139
Research-based knowledge about children's thinking, 94-95
Resolution moves, 145
Restricted-time written test (RTWT), 274, 278, 280
Rosenbloom, Paul, 56
Rote model, inability to move beyond, 3
Russell, Bertrand, 36-37

St. Anselm's ontological proof, 46
Santos, A. M., 65, 66
School mathematics
  vs. inquiry mathematics, 123, 130
  tradition of, 11-13, 120
School tests, 278-279
Schooled populations, understanding of negative numbers in, 66
Schooling, effect of on problem solving, 66
Secondary school curriculum, 86
Self-criticism vs. defensiveness, 26
Self-monitoring, 17, 194
Self-report procedures, 239, 248-250, 255-256
Seneschal, Marjorie, 47
Severity of student errors, 17
Single-attribute conceptualization, 51-54
Situated mathematics, role of, 51
Skepticism, importance of, 26
Small groups
  instruction in, effectiveness of, 19
  problem-solving episode with, 119, 120-123
  work in, importance of, 225
Socioeconomic status (SES), 11
Sociometric interview, 144
Software as part of pedagogy, 91
Solution
  by group consensus, 142
  paths of, 137, 271
Specificity, extreme, 12
Stand and Deliver, 44
Standardized tests
  and assessment, 239
  national, 272-274
Strategies
  children's problem-solving, 95, 183, 247, 248
  coding observed, 256
  valuing the range of, 257
Strickland, Dorothy, 14-16
Student-designed tests, 284-286
Student math
  in elementary school, 215
  in middle school, 215-216
  in senior high/college, 217
"Students-and-professors" problem, 12, 182, 217
Subtraction as a binary operation, 55-57, 58
Swan, Alan, 225
Symbol manipulation, effects of technology on, 84-85
Symbols and meaning, 54-55, 220

Tasks
  for small-group problem solving, 163-165

Tasks *(Continued)*
  real, 263–287
Teach for America, 4
Teacher
  development project for, in mathematics, 141
  direct instruction, 98
  education of and curriculum development, 95
  as facilitator, 143
  roile of the, 103–105, 141, 259
  as thinker, 258–259
  training of, in Cognitively Guided Instruction, 96–115
Teachers
  collaboration of math and art, 47
  decision-making freedom of, 30
  effects of thinking curriculum on, 258–259
  need to prepare for teaching with technology, 86–87
  obligations of, to mathematically talented students, 48
  as professionals, 3–5, 97
Teaching
  decline in mathematics, 83
  to foster growth, 48
  made more difficult, 278
  for mathematical understanding, 119
  modeling and, 259
  negative effect of, 72
  strategies for, derived, 228–229
Technology
  in the classroom, 6, 79–92
  effects of on zero-based curriculum, 84–85
  impact of, 79–92
  "knowing is doing" and, 88–90
  and the mathematics teacher, 86–87
  rapid advancement of, 81
  and research on mathematics education, 87
  right and wrong use of, 90
Temperature as an attribute, 52
Testing and the improvement of learning, 281

Test of Early Mathematics Ability (TEMA), 246–247, 253
Test-oriented teaching, 277
Tests
  designing new formats for, 266–267
  free production, 281
  as part of instructional process, 281, 286
Thinking
  aloud procedure, 248
  assessment, 257
  curriculum in, 237, 257–259
Topics in mathematics, requirements for, 84
Triangles, difficulty recognizing, 216
Truth tables and Wittgenstein, 47
Two-attributes conceptualization, 51–54

Unary operations, 55–56
Understanding
  assessment of children's, 99
  of debts and losses among the unschooled, 65
  mathematics, 79–92
  of signed numbers in everyday life, 74
  students' mathematical thinking, 141
  teaching for mathematical, 119
Urban minority students, programs for, 23–26
Using children's thinking, case study on, 96–114
Using concrete materials to build models, 177

Van Hiele phases of learning, 226, 227
Verb vs. noun knowledge, 21
Videotapes, use of, 17–22, 29, 174, 175, 182
Videotaping students at work, 17–22, 144, 173
Vygotsky, L. S., 136–137, 251

Weil, Simone, 43
Word problems, children's solving of, 95
Written vs. oral representation during problem solving, 66–72

Zero-based curriculum, 79, 82–85
Zone of proximal development, 251

ABW 9222

LIBRARY
LYNDON STATE COLLEGE
LYNDONVILLE, VT 05851